Science Beyond the Classroom Boundaries for 3–7 Year Olds

Science Beyond the Classroom Boundaries for 3–7 Year Olds

Lynne Bianchi and Rosemary Feasey

Illustrations by Simone Hesketh

Open University Press

Open University Press
McGraw-Hill Education
McGraw-Hill House
Shoppenhangers Road
Maidenhead
Berkshire
England
SL6 2QL

email: enquiries@openup.co.uk
world wide web: www.openup.co.uk

and Two Penn Plaza, New York, NY 10121–2289, USA

First published 2011

A catalogue record of this book is available from the British Library

ISBN-13: 978-0-33-524129-3 (pb)   978-0-33-524130-9 (hb)
ISBN-10: 0-33-524129-8 (pb)   0-33-524130-1 (hb)
eISBN: 978-0-33-524131-6

Library of Congress Cataloging-in-Publication Data
CIP data applied for

Typeset by RefineCatch Limited, Bungay, Suffolk
Printed in the UK by CPI Antony Rowe, Chippenham and Eastbourne

Fictitious names of companies, products, people, characters and/or data that may be used herein (in case studies or in examples) are not intended to represent any real individual, company, product or event.

The **McGraw·Hill** Companies

To my new daughter, Claire, love Mummy.

To my dearest friend Alison, love Rosemary

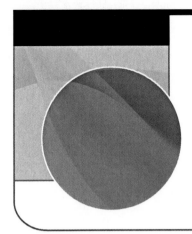

Contents

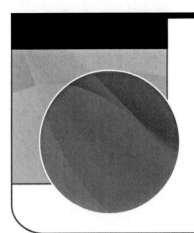

List of figures

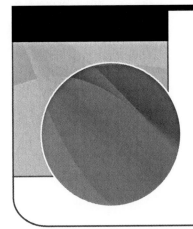

List of tables

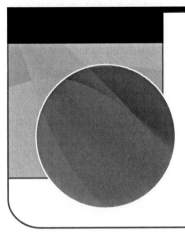

List of illustrations

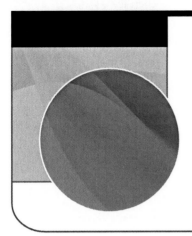

Acknowledgements

We are indebted to the following schools and teachers who have been so generous with their time and in sharing their wonderfully creative practice with us for the purpose of this book. We would also like to thank the many teachers we have met along the way who have been kind enough to allow ideas to be shared. You are all an inspiration.

Bradway Primary School, Sheffield

Bradway Primary School (formerly known as Sir Harold Jackson Primary School) is a large primary school located in the south-west of Sheffield surrounded by beautiful countryside less than a mile away from the border with Derbyshire.

Castleside Primary School, Consett, Co. Durham

Sarah Doneghan (change of name after marriage) is science leader at this small primary school which is tucked away in a housing estate in Castleside, in a small village situated two miles south-west of Consett.

Cheveley Park Primary School, Durham

Sue Harrison was science leader at this primary school on the outskirts of Durham City. The school has a separate Language Resource Base (LRB) which caters for 20 pupils. All pupils in the base have a statement of special educational needs. This means that one in every seven children attending the school has a statement and nearly one in three children are identified as having learning difficulties and/or disabilities.

Fox Hill Primary School, Sheffield

Fox Hill Primary School is an inner-city school with children of mixed ability included an integrated resource, they would describe themselves as welcoming: a friendly school and a fun place to be.

Framwellgate Primary School, Durham

This school is aiming to become a centre of excellence for using the outdoors with all of its pupils from Foundation Stage to Year 6. They are completely rethinking their use of the outdoors for science and other areas of the curriculum and intend to create similar outdoors opportunities to those in the Foundation Stage for every year group.

Gainford CE Primary School, Co. Durham

This is a small primary school with around 90 children in a charming village on the north bank of the River Tees. The school was built in 1857 so space is limited but the school grounds offer lots of potential for work outdoors.

Grenoside Community Primary School, Sheffield

Grenoside is a school with a 330-pupil intake in a new school with grounds ripe for development. The school is on the north side of Sheffield surrounded by a varied landscape of woodland, parks, village and town life.

Monteney Primary School, Sheffield

Monteney is an inner-city school in Sheffield: a lively and fun place for both staff and children to learn and achieve. Pupils enjoy a varied and creative curriculum.

Shaw CE Primary School, Shaw, Wiltshire

This is a small semi-rural school with a mix of new and old buildings and where the school grounds are being creatively developed for science outdoors.

St. Thomas More RC Primary School, Sheffield

This is a small catholic primary school in the north of Sheffield, set in beautiful grounds and ideal for outdoor learning.

Wheatlands Primary School, Redcar, Cleveland

This school is on the southern outskirts of the seaside town of Redcar and has limited room inside but lots of room in the school grounds for doing science.

Woolley Wood Special School, Sheffield

This is a primary school for children with severe learning difficulties and profound and multiple learning difficulties. The school is situated in the north of Sheffield and has a wide catchment area, taking children from $2\frac{1}{2}$ to 12 years old.

Our heartfelt and sincere thanks go to Lynne's dad, who so kindly read through the draft manuscript and corrected our silly sentences and offered suggestions for changes. (Love you Dad – Lx)

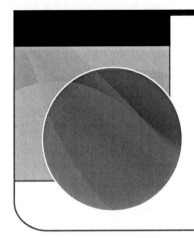

Introduction

Our aim in this book is to offer teachers new ideas for science 'Beyond the Classroom Boundaries', and for some to change their perceptions of where science can be taught. Underpinning this is the resolve to develop children's Personal Capabilities (Bianchi, 2002) so that eventually children are able to choose why and when to take their learning 'Beyond the Classroom Boundaries', on a regular basis. This is not a traditional approach in primary science but a bold step which results in a whole-school approach to moving the science curriculum outdoors into the school grounds.

For the reader it will be important to make clear some of the parameters of this book, by defining two key phrases. First, the term 'early years'; we are identifying this as the period of formal schooling from 3 to 7. Second, the phrase 'Beyond the Classroom Boundaries' refers to the outdoors environment within the school grounds.

Why have we chosen to focus on the school grounds? Quite simply because there is already a great deal of material available to schools from a range of different organizations relating to school visits and the use of the environment beyond the school; however, there is very little material that supports schools in developing and using the school grounds to its full potential in science.

Each chapter in this book is structured into four sections, enabling the reader to consider the aims to be achieved (the goals), what the current situation in schools is (the reality), suggestions for development (the options) and possible ideas for what you will be able to do in your school (the will). This structure draws on the established 'GROW' model which is a technique for problem solving or goal setting, developed in the UK by Graham Alexander, Alan Fine and Sir John Whitmore and is often used for the purpose of coaching. The value of GROW is that it provides an effective, structured methodology which both helps set goals effectively and is a problem-solving process.

Stages of GROW

There are a number of different versions of the GROW model. This version presents one view of the stages but there are others.

Table I.1 The GROW model

G Goal	This is the endpoint, where you want to be. The Goal has to be defined in such a way that it is very clear to the child when you have achieved it
R Reality	This is how far you are away from your Goal. If you were to look at all the steps the child needs to take in order to achieve the Goal, the Reality would be the number of those steps he/she has completed so far
O Options	Sometimes also including an exploration of the obstacles, this stage asks you to identify ways of dealing with them and making progress. These are the Options
W Will or Way Forward	The Options then need to be converted into action steps which will take you to your Goal. These are your 'Wills' or Ways Forward

References

Bianchi, L. (2002) Teachers' experiences of the teaching of Personal Capabilities through the science curriculum, Ph.D. Thesis, Sheffield University.

For more information on the GROW model visit www.en.wikipedia.org/wiki/GROW_model

Transforming science 'Beyond the Classroom Boundaries'

What is the goal?

There have been many years of innovation in primary science education. Surprisingly though, most of this has taken place within the confines of the classroom. What primary science has not yet done with universal success is to step outside the classroom boundaries to use the school grounds for teaching and learning across the science curriculum.

> There is a long history of what is taught there, by whom and how, as well as embedded perceptions about the types of buildings and spaces where these practices traditionally occur. However, there is a need to challenge existing assumptions . . . this requires taking a different approach to imagine a whole range of different possibilities.
>
> (Futurelab, 2008: 12)

As we indicated in the Introduction to this book, for the reader it will be important to make clear some of the parameters of this book by defining two key phrases. First, the term 'early years': we are identifying this as the period of formal schooling from the age of 3 to 7 years. Second, the phrase 'Beyond the Classroom Boundaries' refers to the outdoors environment within the school grounds.

Why have we chosen to focus on the school grounds and not the outdoors generally encompassing different outdoors areas beyond the school grounds that children can visit? Quite simply, because there is already a great deal of material available to schools from a range of different organizations relating to school visits and the use of environments external to the schools. However, there is very little material that supports schools in developing and using the school grounds to their full potential in science.

Throughout this book an equally important element is the development of children's Personal Capabilities (Bianchi, 2002), namely problem solving, creativity, communication, teamwork and self-management. As co-author of this book, Lynne's work has explored how the proactive use of these personal skills and capabilities can help develop children as increasingly scientifically capable and responsible learners, who have:

- an improved understanding of 'how' to learn
- better learning of science

- greater motivation to learn science
- increased engagement in learning science
- improved personal responsibility when learning science.

The basis of this book is the excellent practice to be found in many Foundation Stage settings, since those working in this area have already challenged the idea of where learning takes place and have an open-door policy where children move freely from inside to outside and back. In this book we aim to share ideas to make sure that learning outside in the school grounds has a significant impact on children learning in science up to the age of 7, after which the companion book *Science Beyond the Classroom Boundaries 7–11* supports practice in the later primary years.

In truth, there is an even more fundamental reason for using the school grounds to support primary science. Quite simply, the school grounds are both accessible and free providing the opportunity to make the most of what we have! Our goals should be to support teachers across the early years, to:

- rethink where science learning takes place and to move most science learning 'Beyond the Classroom Boundaries'
- develop an outdoor space that provides a wide range of opportunities for teaching and learning in science and associated Personal Capabilities.

What we hope is that eventually teachers of early years children will automatically plan their science to be outdoors as the norm and indoor science will be more the exception. Robert Brown MSP, Deputy Minister for Education and Young People offered a thought provoking question when he commented:

'We must challenge people to think: Why learn indoors?'

(Brown, 2010)

We think that a very chatty Kelsey (age 5) and her quieter friend Tilly (age 5) would agree with Robert Brown, as their conversation indicates.

Kelsey:	We were putting the car on the track and it lands on water, the car likes having a car wash, doesn't it Tilly?
Kelsey:	Because outside is more fun we get to run around don't we Tilly?
Kelsey:	We like being outside in the rain we have more fun we get our welly bobs on.
Kelsey:	Even if it is chucking it down I don't mind being outside.
Tilly:	I like getting wet best.

What is the reality?

Our work with project schools across the UK has shown that in most primary schools children in the very early years (age 3–5) are most likely to spend between 50 and 90 per cent of their time outdoors, moving freely between the indoors classroom and the external environment. As children move through the early primary years (age 5–7) and then into middle primary years, this time gradually diminishes until, by the time children are in their final year at primary school, the outdoor time averages between 0 and 10 per cent of their science time and is mostly teacher-dominated.

In the early years (age 3–5) a well-planned outdoors environment and the expectation that children will go outside in most weathers is the norm rather than the exception. This changes and children's opportunity for outdoor science diminishes quite dramatically as they get older. Teachers offer a range of reasons, which include:

- time
- the curriculum
- resources
- behaviour
- inspections
- adult supervision
- health and safety
- teacher confidence and ability
- the weather.

None of these is an adequate reason for not continuing the good practice begun in the early years of schooling, and all of them could be overcome. We would then be more likely to develop a science curriculum that made good use of the outdoors and that was new and progressive in moving teaching and learning forward and in offering children teaching and learning that was continuous and incremental and helped them to progress as individuals.

The reality of most classes for 3–5 year olds is, as we have already stated, that outdoors activity is the norm. However, one of the issues for primary science is the quality of the activity. This often lacks the focus on science, failing to demonstrate to children how they are engaged in science, talking about scientists and explaining to children that they are using special science words. Doing these things would provide a firm basis for ongoing development and an appreciation of the wealth of happenings around them. As teachers we do not hesitate to tell them that they are doing literacy, maths, art or PE. Equally, we firmly believe that children should know that they are doing science – and celebrate this area of the curriculum – otherwise, how can children develop their understanding of what science is and what it means to work like a scientist if they are not engaged from an early age in the language of science and meet scientists from the workplace?

What are the options?

As children move through the early years, it is less likely that their experience will mirror their first year in school with the wide range of activities and access to the outdoors. Thankfully many schools are beginning to rethink this situation to ensure that the initial experiences are continued. They are beginning to take down both the physical and mental walls across the years, and as a consequence schools are moving to a less formal regime. Different schools have approached this in different ways depending on the geography of their school grounds and buildings:

1 Where settings and classrooms are adjacent they are merging the first year of school with later early years classes.

2 If the geography of the school does not allow year groups to merge, some schools are creating outdoor areas modelled on the first year of school directly outside classrooms for 6–7 year olds. For example, La Houguette Primary School on Guernsey has built a fence around the area immediately outside each early years classroom and begun to develop these areas in a similar way to their Foundation Class, by introducing sand, water, large toys, role play and raised beds for planting.

3 Where settings and classrooms are further apart, older children have been timetabled to join younger children outdoors.

4 In a minority of schools where the school landscape and geography of the buildings does not allow for children to move between the classroom and outdoors, then timetabled regular periods outdoors are planned into the school week.

Our options are quite clear: we need to move primary science 'Beyond the Classroom Boundaries' and to use the school grounds across early years so that children develop and apply their Personal Capabilities when thinking and working in science. The alternative is to maintain the status quo that, in our opinion, would be to restrict the opportunities, and to limit the potential for learners, denying them a creative curriculum and good quality primary science and removing the opportunities for children to progress in their science 'Beyond the Classroom Boundaries'.

What will you do?

Here are some suggestions for where to begin:

1 Develop a holistic early years science policy which focuses on children's learning in science being, in the first instance, outdoors, by asking the question 'Why are the children indoors for this science experience?'

2 Review which elements of the science curriculum could be taken outdoors, if you find it is weighted towards the natural sciences and teacher-directed observations, then move experiences towards the physical and material science areas and make positive changes in planning towards a more balanced diet of:

- spontaneous play/exploration
- planned play/exploration
- problem solving.

3 Talk science with staff, in terms of curriculum planning, and to children.

4 Talk personal skills to staff, in terms of curriculum planning, and to children: team-work, communication, self-management, problem solving and creativity.

5 Bring scientists, and people who use science into the early years, as role models to work alongside staff, children and parents.

6 Continue very early years practice where adults take time to sit and wait to see if children will provide their own solutions to questions and problems.

7 Ensure continuity across the early years in terms of children being the leaders of their own learning, which therefore requires teachers to adopt responsive planning for science 'Beyond the Classroom Boundaries'.

Practical ideas on senses for transforming learning 'Beyond the Classroom Boundaries'

Senses science day or week 'Beyond the Classroom Boundaries'

If you think that you could offer more science experiences for children 'Beyond the Classroom Boundaries' but you are not quite sure where to start then why not just begin in a small way by dedicating 'special time' to science outdoors. How about having a 'Senses Week'? Here are some ideas to get you going.

Create a 'Senses' area
Smell
Forget the smell pots, bring in smelly flower pots: for example, lavender, sage, thyme, curry plant, chocolate cosmos flowers, rosemary, parsley, stocks, lemon balm and different kinds of mint – for example, apple mint and spearmint. Many of these parents and members of staff might donate; alternatively, contact a local garden centre and ask them to donate plants. Different centres might each donate one or two and therefore your collection will grow. All of these plants offer wonderful aromas – some of which will require children to rub the leaves in between their fingers thereby offering senses experiences of both smell and touch. Even better would be to harvest some of these to use to cook: for example, mint in couscous, thyme and marjoram in a bolognaise sauce or as a tea drink.

Taste
Normally we would not encourage children to taste anything in the environment. However a Senses Week enables us to train children in this aspect of health and safety by insisting that they are only allowed to taste anything outside with adult

7

supervision. So children should be able to taste crops from the garden and, where appropriate and possible, as soon after they have been harvested. Try to widen the crops grown in your school grounds – for example, you could include:

apples	blackberries (thornless)	blueberries	broad beans
carrots	courgettes	cucumber	french beans
garlic	leeks	lettuces	onions
parsnips	pears	peas	potatoes
radish	runner beans	spinach	spring onions
strawberries	sweetcorn	tomatoes	

Instead of taking children indoors, take a bowl of water outside to wash the vegetables; cut them and allow children to observe before and after washing, before and after cutting: smells and tastes. Let children taste pairs of vegetables: for example, different lettuces, tomatoes and spring onions, apples and blueberries, so that they can think about similarities and differences in colour, size, shape, texture, smell and taste, and also to consider combinations of flavours.

Sight

We're going on a Smart Hunt: This activity is based on Smart Hunt which is one of the activities in the Smart Science resource from Sheffield Hallam University. More information about this can be found online (at www.smart-science.co.uk). This could take place at a set time during a day or children could be offered different Smart Hunts related to a variety of science concepts on each day of a Senses Week. Depending on children's abilities they could be given:

- a written list
- photographs to match
- bag of objects to match by sight
- feely box to match the texture and shape
- an ipod, with instructions of what something looks like e.g. size, shape, texture and colour.

Children could be given the same or different things to find on their Smart Hunt, using the sense of sight, such as:

- invertebrates
- leaves or plants
- objects of a certain colour
- six objects that are different shades of green
- objects that are round

- science equipment that has been 'planted' around the area which has particular functions e.g. something to make things look bigger, something to put plants in, etc.

1 Organize the children into teams of between three and six. Tell them that they are going on a Smart Hunt to find certain things in the school grounds using their sense of sight.

2 Give out the job Role Badges and the Smart Hunt Cards. Ask them to discuss and reach agreement about who will do what.

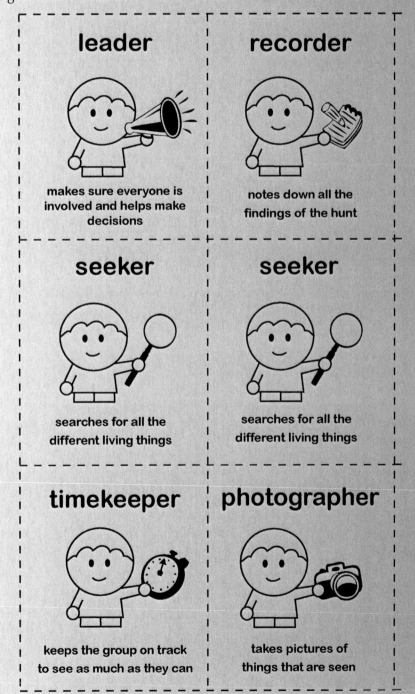

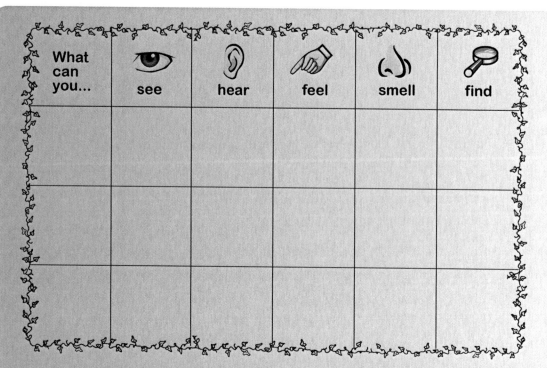

What can you...	see	hear	feel	smell	find

1.1 *Smart Hunt Cards*

3   Give instructions as to where the children may venture during their Smart Hunt, the timings they should be aware of and the health and safety issues in your location.

4   Stagger the start of the hunt by allocating particular groups particular 'Hunt Hotspots' (areas of the school grounds). When arriving at the Hotspot they should chant their Smart Hunt rhyme:

We're going on a Smart Hunt.
We're going to catch a big one. [or what the children are searching for]
What a beautiful day!
We're not scared.

Uh-uh! Gravel!
Stoney, hard gravel.
We can't go over it.
We can't go under it.
Oh no! We've got to learn from it!

They should be given up to 10 minutes in each Hunt Hotspot before moving on.

5   Indicate that as a team they should get on together to complete as much of their Hunt Card as possible, by exploring and examining the area using the equipment provided.

Texture Barefoot Trail

Barefoot Trails are great fun and often found in outdoor adventure playgrounds. The concept is very simple: that is, to give children 'Barefoot' experiences in relation to learning about the senses and texture, and so extending children's understanding of how we feel, and that we can feel with our feet as well as our hands.

Of course, you would need to consider health and safety: for example, making sure that children cannot get splinters, there are no sharp edges and that surfaces do not become slippery and therefore dangerous.

So, how could you set up your own Barefoot Trail, you could use:

- a series of paddling pools across the grass or concrete
- a series of shallow plastic bowls set at different intervals either in a straight line or snaking around the area.

There are many things that could be used to offer a range of experiences of different textures, which could be organized into groups and in sequence, so that children end up in a substance that helps to clean their feet or ensure that they walk through warm water and then onto a towel. Table 1.1 contains suggestions.

Table 1.1 Textures for Barefoot Trail

Sequence	Dry	Wet	Hard	Soft	Squelchy
1	Leaves	Wet sand	Pebbles	Cotton wool	Jelly
2	Sand	Clay	Wood	Silky fabrics	Porridge
3	Straw	Mud	Linoleum	Natural sheep's wool	Bath gel
4	Clean compost (for soil)	Water	School ground surface	Bubble wrap	Gloop

If we take a leaf from Michael Rosen's book, *Going on a Bear Hunt*, we could sing a similar song, 'We're going on a Barefoot Trail' using the words the children use to describe the textures, and of course those introduced by the adult to extend and enrich children's scientific language, such as:

hard	soft	silky	smooth	furry
bobbly	sticky	squashy	runny	gloopy
spongy	slippery	scratchy	bumpy	downy

This activity is excellent to engage children in learning about materials and textures, as well as similarities and differences, which of course means that children will be introduced to comparative terms in science such as same as, different, similar, like, as well as 'er' endings: for example, harder and softer.

As well as having lots of fun and there being excitement and motivation for this type of learning, it also requires that children take turns, listen and communicate using appropriate language. Indeed it may also require a degree of self-control for some children, which needs to be encouraged through explanation, example and reward. Instead of focusing on such an array of personal skills it is suggested that teachers consider which are most relevant for the age and stage of the children. Ask them to take turns when working in a team, or to focus on using appropriate language, or to take charge of themselves. During the early years a specific objective for Personal Capabilities such as these is more than enough, especially when linked to such rich experiences. Use verbal review during mini-plenaries or at the end of activity to discuss with the children how they learnt and acted on the task, and ask some to model good behaviours. Reward positive effort rather than excellence. In practice we do not all get it right the first time, but investing effort in reaching objectives is highly motivational and builds positive mindsets.

Elephant Ears

Sound – Elephant Ears
This activity is based on work carried out with a class of 6–7 year olds at Castleside Primary School in Consett; it is a great activity, so have a go.

The children were engaged in sound activities and talked about their ears and hearing sounds. The teacher suggested that if they had really big ears perhaps they would hear more when they were out and about in the school grounds. So the children designed and made their own Elephant Ears, and wore them on a walk around the school where their task was to listen very carefully so that they could hear lots of different sounds. They were

1.2 Elephant Ears

given a clipboard with paper and pencil (which made them feel like real scientists) and they wrote or drew whatever sounds they could hear. When they had walked around the school grounds they then found a space and sat on their own and listened for a few minutes, then they joined a friend and listened together, after which the children talked animatedly to each other about the sounds that they had heard.

The result, as the teacher explained, was quite amazing: the children were delighted to wear their own Elephant Ears which helped them get in role. The children understood that the Big Ears were designed to pick up even the quietest sound, and they listened intently, and would you believe that with the aid of their Elephant Ears the children picked up far more sounds than previous classes had done!

Of course, Elephant Ears were to be seen on heads even after the activity had formally ended.

References

Bianchi, L. (2002) Teachers' experiences of the teaching of Personal Capabilities through the science curriculum, Ph.D. thesis, Sheffield Hallam University.

Brown, R. (2010) Available online at www.eriding.net/edu_visits/learning.shtml (accessed 25 June 2010).

Futurelab (2008) *Reimagining Outdoor Learning Spaces: Primary Capital*, Co-design and Educational Transformation. Bristol: Futurelab.

CHAPTER 2

Identifying the potential for science 'Beyond the Classroom Boundaries'

What is the goal?

In this chapter we attempt to show how easy it is to:

- identify how the school grounds are currently used for science
- identify the potential of the school grounds for science
- involve children in developing the school grounds to support science
- develop an action plan to achieve the redesign of the school grounds.

A key issue here is the involvement of the whole school, including the children, since there is 'evidence to suggest that adults gain a far better understanding of children's capabilities and interests and are often surprised by the skills, aptitudes and resourcefulness of children involved in co-design projects' (Futurelab, 2008: 21). Children do not have the conceptual barriers of time, funding or feasibility that can often inhibit the imagination of teachers. Thus, since changes are directed at children using the school grounds, it would seem obvious that children should be involved in the process. This is regardless of whether yours is through a primary school or not, continuity and progression in science will depend on science indoors and outdoors being seen holistically.

What is the reality?

One of the first steps is to identify how the school grounds are currently used for science. Teachers who have worked with this project began by mapping out their school grounds and annotating their maps using Post-it notes or coloured pens; they identified areas that teachers already used for science, indicating who, when, why and how the areas were used.

Your reality – the school grounds

To begin the process it is useful as individual members of staff and as a whole school to reflect on and answer the following questions:

- How do the school grounds already support science?

- What limitations/issues are there in relation to science?

- What is the potential of the school grounds for the different areas of the science curriculum?

- What would need to be changed to achieve the potential for science 'Beyond the Classroom Boundaries'?

- What equipment do we already have that can be used for science outdoors?

- What equipment for outdoors do we think would enhance children's experiences in science?

- What funding do we have available? What funding do we need? How will we make up any deficit?

- In what ways do the school grounds encourage independent learning?

Your reality – the curriculum

Auditing the school grounds is a good place to start, but the potential of science 'Beyond the Classroom Boundaries' will be fully realized when planning to use the outdoors for science has become second nature and there is free movement between indoor and outdoors.

Figure 2.1 will help you and staff to move forward with planning for science, and the following set of questions provide focal points for discussion:

- What percentage of children's experiences in science takes place 'Beyond the Classroom Boundaries'?

- How can you improve on that percentage?

- Which outdoor areas are used most for science? How can you celebrate this? Is there progression from one year group to the next in children's learning.

- Which outdoor areas are underused in science and why? How can you overcome this?

- Which aspects of science could be developed in which area of the grounds? How?

- If the use of the school grounds for science is seasonal favouring the warmer months, which type of topics would work outside during colder months?

- Are the school grounds used more frequently by younger year groups? If so what will you do?

- Are science activities outdoors more likely to be linked to natural science, with few activities linked to the material and physical sciences? How can you ensure that children's learning is balanced across the science curriculum?

- Are activities in science usually teacher-initiated and -controlled? Are they activities where children mainly observe and collect? How can you move learning towards being independent, explorative or investigative, where children take the initiative and are in charge?

- Do children take part in longitudinal studies (e.g. across seasons)?

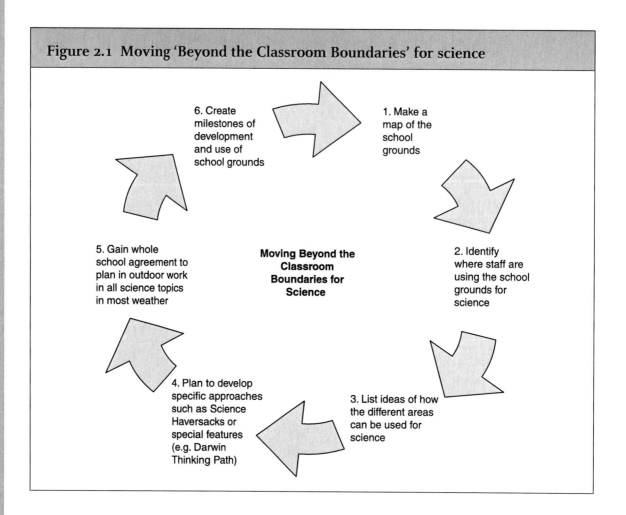

Figure 2.1 Moving 'Beyond the Classroom Boundaries' for science

6. Create milestones of development and use of school grounds

1. Make a map of the school grounds

5. Gain whole school agreement to plan in outdoor work in all science topics in most weather

Moving Beyond the Classroom Boundaries for Science

2. Identify where staff are using the school grounds for science

4. Plan to develop specific approaches such as Science Haversacks or special features (e.g. Darwin Thinking Path)

3. List ideas of how the different areas can be used for science

The answers to these questions will help to ensure that the potential for planning and learning for science 'Beyond the Classroom Boundaries' is realized within the school grounds.

What are the options?

Collaboration in planning

Across the early years, staff should also consider how the outdoors is organized in terms of activity areas that can support science. While settings for 3–5 year olds usually have a range of areas and activities which have potential for science, opportunities are often more limited for 5–7 year olds. So there is a question of parity of provision across the early years and how do we ensure that this provision is progressive. The examples in Table 2.1 illustrate how early years teachers can collaborate to make sure that all children have access to outdoor areas such as sand, water and construction (areas such a planting and role play are in other chapters in this book) to ensure that the demands of the activity in terms of science and Personal Capabilities progress as children move through the early years of schooling.

Table 2.1 Parity and progression across the early years

<table>
<tr><td colspan="2" align="center">Area: Water</td></tr>
<tr><td colspan="2">

Science
- Observation, noting changes, predicting, making connections
- Water changes shape, can be poured, can be absorbed, water can disappear (evaporate), water can make the colour of objects look different (e.g. bricks and slate)

Personal Capabilities
- Creativity: to ask questions about the world around us
- Self-Management: to keep track and monitor changes around us

</td></tr>
<tr><td>Activities for 3–5 year olds</td><td>Activities for 5–7 year olds</td></tr>
<tr><td>

Bucket of water and collection of brushes of all sizes and dimensions plus rollers and paint trays, sponges, etc.: 'paint' anywhere and everything using a two-step ladder so that children can reach high up; an all-in-one waterproof 'decorator's suit' for role play.

What was the wall like before?

What happened when you painted the wall? (e.g. wall went darker, wet, there were drips)?

Will the water still be on the wall after we have had our snacktime?

Encourage children to use basic question stem cards to come up with a range of questions they would like to explore outdoors. Provide question starts, such as what, when, how, if, what if . . . to help them develop a range of questions. Consider those that can be followed through.

</td><td>

Creating puddles in different places in the school grounds. Measuring the length and breadth of the puddles each day: do they get bigger or smaller? What is the weather like?

What do you predict will happen to the puddles? What makes you think that?

Take photographs

Draw with chalk on the playground where you think the water from the puddle goes.

Keeping an eye on the changes around and the influence of weather, seasons and us on the environment begins to assist the children in acknowledging the changing world around us. Encourage them to develop the skills of keeping track of changes, using simple charts, photographs, drawings, models, etc. And where possible encourage them to monitor the range of change. Including the use of ICT for this purpose is ideal.

</td></tr>
<tr><td colspan="2" align="center">Area: Sand</td></tr>
<tr><td colspan="2">

Science
- Sand has weight
- Sand poured onto or into some materials can make them stretch
- Sand takes the shape of its container
- Measuring, simple graphs

</td></tr>
</table>

(Continued Overleaf)

17

Table 2.1 Continued

Personal Capabilities
o Communication: to share ideas and opinions

Activities for 3–5 year olds	Activities for 5–7 year olds
Collect empty plastic food and drinks containers. Punch holes in them for children to observe how the sand flows through. Make sure holes are different sizes to allow for a drizzle to a rapid pouring of sand. Give children a marker pen to mark on a container where they would like the adult to punch holes. Begin with questions, why are you putting the hole there? What do you think will happen? Later introduce other containers e.g. cornflake boxes, tubes, plastic tubing. Use a washing-up liquid bottle with the base cut off and give it a string handle so children can let it swing and watch the patterns the sand makes as it falls into the tray. Encouraging children to talk freely about what they see and to share ideas and opinions about why this is happening are basic skills for early years development. Use such science activities to encourage children to begin to understand the difference between ideas and opinions.	Which sock holds the most sand? This is an easily resourced activity which can offer a range of outcomes. Socks are made from different fabrics and stretch to varying lengths, some almost as tall as the child. Get children to put the sock onto squared paper, stretch it themselves and make a mark, then put sand in it and see if it stretches further, mark the paper with comparative length. When comparing outcomes from these activities, develop the language to encourage detailed descriptions–longer, shorter, heavier, stretchier, elastic, etc. Ask children to go beyond basic observations and to share their ideas as to why they think particular things are happening. Accept and reward tangible answers and creative considerations.

Area: Construction

Science
o Materials and their properties
o Language of materials and properties (e.g. hard, soft, transparent, waterproof, warm)
o Testing materials, measurement, prediction, results, recording, conclusions

Personal Capabilities
o Teamwork: to be a helper and work well with other people

Activities for 3–5 year olds	Activities for 5–7 year olds
Make a den. Give children access to a wide range of materials, allow children to make and change their den over time. Focus on the materials that they need to use and early understanding of their properties, offer children card, wood and plastic crates or boxes, tarpaulins, blankets, strong tape (e.g. gaffer tape). Developing social skills is another key personal capability at this age. Beginning with partner or Buddy, encourage children to work together, taking turns, sharing out jobs, talking about ideas, and reaching decisions and agreements together.	Design and make a tent – test the materials to be used for the camping tent and sleeping bag. Give children access to cotton sheets, canvas, plastic sheets, net curtains for water and windproof. Children will also need to test materials for a groundsheet and sleeping bag. Provide children with access to a range of items to help construct the frame for their tent: e.g. washing line (hand material over, large sticks (make a wigwam). The use of team role cards for groups of three or four children are useful means to begin to provide some structure and assistance to working together. Begin with roles such as 'organizer', 'reporter' and 'maker'.

What would we like in the school grounds to support science?

In the north-east of England people have a saying 'Shy bairns get nowt!' and a very useful saying it is. If we take a conservative approach to the school grounds we will never help children to realize their potential in primary science. So the most important thing to do at this point is to momentarily put aside all issues relating to the outdoors, such as health and safety, money, staffing, challenging behaviour, and free your imagination to create, on paper, your ideal school grounds for science. Teachers and children who tried this found it very liberating and it led to a range of suggestions, such as those in Table 2.2.

Kay Coverdale, who led the thinking in her school, developed some quick, easily achievable ideas, such as creating Outdoor Science Resource Boxes. Others, such as developing areas directly outside each classroom required long-term planning and some core expenditure. However it is important to remember that the movement in teaching and learning at 5–7 years old is towards mirroring some of the very early years (age 3–5) practices where the independent movement of children between indoors and outdoors is well established.

Table 2.2 Teacher suggestions for the ideal outdoor environment

We would like	Making it real
Outdoor resource kits	Outdoor boxes containing, for example, binoculars, pooters, paper, pencils
Science area outside each classroom	Perspex roof outside classrooms for shaded area where children can work
Work stations for children to carry out practical activities	Tables to go outside each classroom permanently
Planting area outside each classroom	Plant pots, tyres, unusual containers, wooden raised beds, railway sleepers, for children to plant, care for and carry out investigations
Darwin Thinking Path	Designated walk around the perimeter of the school grounds with arrows, words, animal sculptures placed at points to denote the walk
Boards for writing	White- or chalk boards for children to write or draw on, use as notice or question boards and observation boards
Hide for birdwatching	This could be a tent donated by parents, or a camouflaged hide from science resource catalogues, it could be a purpose built shelter (by friends of the school) with seating, a board for logging observations, identification posters, etc.
Increase use of ICT outdoors	Cameras for children to use outdoors whenever they choose to do so
Bird box and webcam	Sited in school grounds and linked computer in position where children can access it inside the school

Changing the outdoor water area

At Castleside Primary School, teachers of 3–5 and 5–7 year olds decided to reconsider some of the areas that they had for many years taken for granted, such as water outdoors. After reflecting on a series of questions they decided that they would like to change the water area which was fairly basic with a water tray and toys, to make it more stimulating and challenging for children. Ideas for doing this are shown in Table 2.3

Table 2.3 Changing the outdoor water area

Reasons for change	Suggestions

To make the area more stimulating with a wider range of activities, to suit all ages from 3 to 7, so that water experiences are continued throughout the early years	*Focus on specific science outcomes*
	Forces
	Sail boats and fans or straws, plastic straw poles in the water for slalom races – changing direction.
	    Floating and sinking, sponges, partially filled lemonade bottles, sealed transparent polythene bags with different amounts of water inside, grapefruits – peeled and unpeeled.
	    Squirting water using squeezy bottles, syringes, pipettes, turkey basters, water pistols. Who can squirt the water the furthest? Heighest? Hit the target and score points.

To support problem solving	Change the problem regularly to encourage the transfer of problem-solving skills such as prediction, solution finding, analysis and review, for example:
	    Our bucket has a hole in it which is the best way to fill/ empty the water tray?
	    Make a raft to go across the crocodile-infested river, that carries X number of weights or washers.
	    Make a boat to carry a gingerbread man across the river (give the children real gingerbread men); he must not get wet or he will go soggy.

To offer language support to children and adults working alongside groups	A mural is painted on the brick wall behind the water tray with pictures to suggest what children might do: water-related pictures such as water drops, umbrella, clouds. Key water-related words are also included; see the list in the practical ideas section at the end of this chapter.
To encourage and develop responsibility	Hooks for items on the wall with silhouttes to indicate what goes where, net bags for toys, boxes with photographs of different water experience items. Children choose a blue bib, or a bib that says 'Water' to access the water area.

If we engage teachers, pupils and the local community in developing the science potential of the school grounds, then we will be encouraging engagement, ownership and responsibility where children are more likely to:

- use the facilities
- apply science ideas and skills
- be interested in their environment

- care for their environment
- use it to its full potential
- develop personal self-esteem
- develop pride in the school grounds.

Who can support us?

Schools involved with the project soon realized that once they had created their 'We would like list', the next step was to translate this into reality and to do so in the most efficient and economic way. It would be naive to suggest that moving a school towards using the outdoors for science does not require some funding, but teachers recognized that they could request support from a range of people and agencies in the local community to achieve their goals in a cost-effective way.

If we look at Kay's wish list, it becomes obvious that this cannot be achieved within a short period, not only for financial reasons but also because resourcing, planning and embedding takes time. So Kay needed to write an 'Outdoor Science Action Plan', and to consider:

- What would make an immediate impact and encourage children and staff to engage in science 'Beyond the Classroom Boundaries'?

- What could be achieved rapidly and would be used regularly by each year group?

Table 2.4 Action plan for the first year of changing the school outdoor environment

Immediate action	Science Outdoor Boxes – equipment gathered from science resources, science resource budget allocation used for additional equipment.	Spare tables from around the school placed outside each classroom.
Term 1	Planting area outside each classroom: plant pots, tyres, unusual containers, wooden raised beds, railway sleepers, etc. (donations from local community).	Darwin Thinking Path developed. White and chalk boards in each outdoor area. Outdoor texture boards for 3–5 year olds fixed to wall.
Term 2	Bird box and webcam negotiated from ICT and science budget.	Hide to be constructed by parent volunteer group.
Term 3	Digital weather station – used by 6–7 year olds mentored by 9–11 year olds.	Science area outside each classroom – perspex roof for shade over work stations outside ready for new school year. Capital expenditure – also to be used across the curriculum.

- Where items needed to be purchased, what could be made available from the current year's budget?
- What could be sourced from the local community?

What will you do?

A key starting point is to audit the outdoors for its potential to support teaching and learning in science, then to consider your current science curriculum and planning and how you can increase children's exposure to science 'Beyond the Classroom Boundaries'. The next stage will be to consider what can be accomplished by teachers and children immediately, and allocate jobs and responsibilities to different people and year groups across the school. Whatever you delegate must be easy and achievable so that everyone involved is able to feel positive from an early stage in the process. This will give teacher and children confidence to further explore science outside. Remember that this is as much about developing children's personal skills of teamwork, problem solving, creativity and self-management as it is about learning science – that is, the process of learning needs to be considered just as much as the science itself.

As your plans begin to crystallize you will recognize which aspects could benefit from exploring external links to find out which organizations and individuals could support the development of 'Science Beyond the Classroom Boundaries'. Do consider engaging children in this process by asking them to write letters or create short video clips, to inform the audience about the children's ideas for their school grounds. It can often be more beneficial to focus on people's skills and knowledge to support different aspects of developing science 'Beyond the Classroom Boundaries', rather than asking for donations of money.

Finally, consider how you and the children will communicate all of the exciting ideas and events to parents and friends of the school including school governors. This is particularly important for parents, so that they understand the educational value of working in science 'Beyond the Classroom Boundaries' in terms of science skills and knowledge, as well as Personal Capabilities.

Practical ideas for exploring water 'Beyond the Classroom Boundaries'

Revamp your outdoor water area

Do you have an outdoor area where children are engaged in the exploration of water and other liquids? If you do, then take a good look at it. If you do not have one, now is the time to create it. How about trying out some of these ideas? Many areas have either a wall or a fence as the backdrop that could be ideal for creating a water mural, which might be a:

- pond with ducks, swans, fish leaping, boats sailing
- seashore scene with rock pools, children with nets searching for fish and crabs

2.1 Water area redesigned

- clouds with rain, umbrella, children in waterproof clothing jumping in and out of puddles.

Whatever the scene, it could be the result of children working with an artist in residence, local artist, artistic teaching assistant or parent. So having created a visual backdrop with lots of ideas for children to try for themselves, do not forget to add vocabulary on the mural or on the perimeter. The language provides opportunities for adults working with children to point to words, play games to find words, help children to make connections between what they are doing and the shapes of words, initial letters and so on. Here are just some suggestions for words that might be painted onto the mural:

boat	cascade	drizzle	drops	dry	flow
force	lakes	pitter patter	pour	puddles	push
rain	raindrops	sail	splash	squirt	umbrella
water	waterfalls	wet			

Question words could also be placed around the mural, to encourage children to ask questions and develop key questioning skills, which is fundamental for creativity, problem solving and communication:

what	when	how
could	can	what if
where	if	who

So already the area looks different, and now the question is how to organize the area. Will it be water toys: piled in a box or

- linked to topic themes to offer a wide range of experiences over the school year (e.g. boats, pirates)?
- linked to science topics such as forces, floating, sinking?
- organized so that children can see and access (and return) the contents of the boxes or net bags independently of adults?

For those just coming to terms with science 'Beyond the Classroom Boundaries' these will be important starting points for discussion when organizing resources for outdoors.

So what kind of science-based water activities could go on in this great area outdoors?

Water cascades

One could purchase a 'water cascade' or provide five or six pieces of guttering, PE stands, plastic milk crates and so on, thereby allowing the children to create their own water cascade.

The science ideas and vocabulary that we would like to develop with children are:

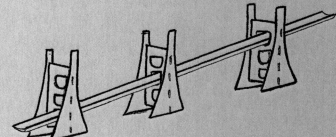

2.2 *Water cascade*

- water is a liquid
- water takes the shape of its container
- running water is a force that can make things move and change direction
- water has a push-up force which we can sometimes feel if we push things down into it.

There are many ways in which the teacher or adult working with children can help to change and progress children's experiences and learning by using the water cascade, for example:

- send water or a more viscous liquid, such as a soap flakes and water mix, down the cascade
- change the angle and direction of the cascade to change the movement of the water

- change the cascade to pour water into a container or to hit a target; perhaps hit a plastic skittle so it will be pushed over
- change the amount of water and increase or decrease the force of the water
- send items down guttering with or without water
- role-play 'Incy Wincy Spider' using a drainpipe and a piece of guttering, plastic spiders and watering cans
- 'Beat the Timer': for example, 10 seconds for a boat to get down to the end of the guttering, or for an object to reach the end of the cascade, particularly for more able children.

Allow children access to cameras or flip cameras so that they can create a 'Photo Book' of what they did and what they found. This will support their observations and reflections, encouraging them to talk to each other about their experiences and achievements.

Bubbles

Bubbles intrigue children and offer rich opportunities for exploration in science where the science ideas and vocabulary that we would like to develop with children are:

MAKING BUBBLES INSIDE A BUBBLE

- water is a liquid
- bubbles are made from liquid
- bubbles can change their shape and size but always return to make a sphere
- we can see rainbows in bubbles
- we can see reflections in bubbles
- bubbles do not like dry things, they burst
- bubbles like wet things and so you can wet your finger and put it through a bubble.

STRING LOOP BUBBLE MAKER.

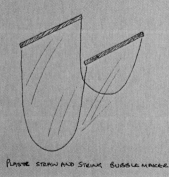

PLASTIC STRAW AND STRING BUBBLE MAKER.

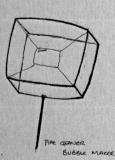

PIPE CLEANER
BUBBLE MAKER.

Below are lots of different items that children can use to make bubbles. A recommended bubble mixture is:

900 ml water
120 ml washing-up liquid
6 tsp. glycerine

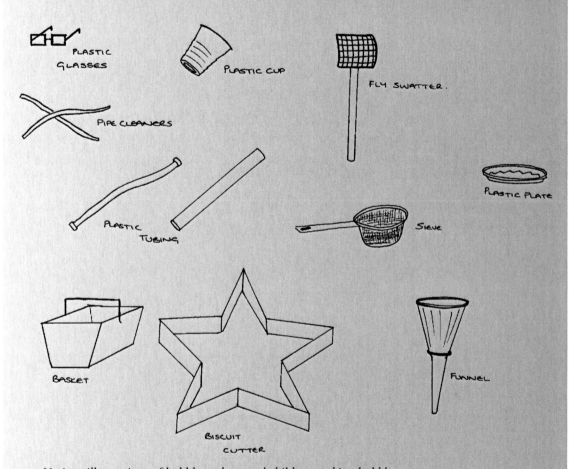

2.3 *Various illustrations of bubble makers and children making bubbles*

If you are making this for a giant bubble (see Illustration 2.4) in a children's paddling pool, you will probably need three lots of this mixture.

Give children opportunities to explore different bubble makers and to observe each other and teach each other how to create bubbles in different ways.

Think about challenging children to solve a problem, such as the one presented in the letter below. Providing contexts for learning focuses and engages children and provides them with a purpose for undertaking an activity and an audience to communicate their findings. Consider the range of different ways, apart from written text, by which they can communicate; for example, Easispeak microphones, Photo Books or a big book.

Bubbles 'R' Us
Unit 7
Durham Industrial Estate
Durham
DH1 1BM

Dear Class 14,

At Bubbles 'R' Us we want to make a new exciting product which children will love to use. We would like to be able to sell the best bubble solution in the world but we need your help.

We would like you to explore and investigate this question:

'Which is the best bubble solution?'

When you have finished your work, our bubble scientists here at Bubbles 'R' Us will need to know how you carried out your tests and your results.

We are very excited about being able to manufacture the best bubble solution so that millions of children around the world can enjoy blowing the most fabulous bubbles.

Yours sincerely

Tina Wand

Director Bubbles 'R' Us

Sarah from Castleside tried this activity with her children; it was amazing. She made up three lots of the bubble mixture the day before and then poured it into a child's paddling pool. Using a PE hoop she created a huge bubble, much to the children's delight. They were however awestruck when she asked one of the children to stand in the paddling pool and then made a bubble that enclosed the child. The class was even more impressed when Sarah then made a bubble with two children inside.

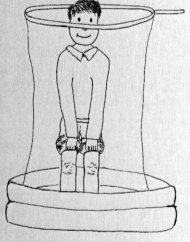

2.4 Boy in a giant bubble

Reference

Futurelab (2008) *Reimagining Outdoor Learning Spaces: Primary Capital, Co-design and Educational Transformation.* Bristol: Futurelab.

CHAPTER

3

Personal Capabilities and science 'Beyond the Classroom Boundaries'

What is the goal?

We have already suggested that in early years teaching the outdoors can be considered as two main areas: the first being an area adjacent to the classroom which is seen as an extension of the indoor area, and the second could be any other area further away from the classroom. Whichever area is being used, the philosophy behind it will mirror good early years pedagogy, where different areas are developed to encourage a range of thinking and learning in science. We would expect to see areas where children can explore, role play, sit and talk, and even hide in a den, as all provide opportunities for science learning and self-discovery. Children should also have access to explore larger physical spaces, where they can develop their understanding of science ideas in relation to large objects: for example, trees and large play equipment. Linked to this, young children should be engaged in science projects that are longitudinal: for instance, observing changes over time, or making something that takes multiple steps.

Unless it is underpinned by the development of Personal Capabilities children's conceptual and skill development in science is somewhat compromised. Children thinking and working scientifically requires a range of Personal Capabilities from working in a team and understanding how to manage a task, to communicating their science. The idea is not new to early years; indeed, personal skill development is the basis of teaching and learning methods and it is well accepted that cognitive and social learning are inseparable from subject knowledge. In this chapter we will build on this early years approach by illustrating that when science is taken 'Beyond the Classroom Boundaries' it is linked to the development of children's personal skills and capabilities.

The Smart Science materials (Bianchi and Barnett, 2006) developed by the Centre for Science Education (CSE) at Sheffield Hallam University explores how Personal Capabilities can be embedded in the science curriculum (Bianchi, 2002). The goal is to enhance children's learning in primary science (and also beyond) by underpinning their experiences and their development of knowledge and understanding, tailoring learning to be as much about 'how' they learn as 'what they' learn. Children can become self-aware, responsive to their own strengths and eventually increasingly 'personally literate'.

As you progress through this chapter and explore activities you will note that attention is paid to developing the language and underpinning knowledge and understanding for a range of skills and capabilities. This is crucial in developing Personal Capabilities, and therefore it follows that it is an important element of children working 'Beyond the Classroom Boundaries' in science.

The ultimate goal – Personal Literacy

Many frameworks have offered skills or competency sets that can be associated with developing children's personal responsibility and independent learning skills (see Lucas and Claxton, 2009). For the purpose of this book five Personal Capabilities will be addressed:

1 Self-management – taking charge of your own learning.
2 Teamwork – working well in groups and teams.
3 Creativity – coming up with and sharing new or unusual ideas.
4 Problem solving – analysing problems and developing strategies and solutions.
5 Communication – speaking, listening and sharing feelings with meaning.

These capabilities have often been considered as endpoints in themselves in that once children become more aware and able to communicate, or manage themselves, they have achieved the required outcomes. These first steps are the foundations of a development that leads to the greater issue of 'Personal Literacy', a term synonymous with being an independent learner.

Personal Literacy defines a range of abilities, from children having the knowledge and understanding of personal skills and capabilities, to being able to speak with confidence about their own personal development. Children developing Personal Literacy can begin to articulate the particular aspects of a skill that they feel is relevant to them and why. They also develop a strong sense of what they can do to help improve their abilities, and many are able to think about where they could find support. Eventually they will have the self-confidence to demonstrate their skills to others, receiving and giving feedback where appropriate. Early years teachers are crucial in laying these foundations, and in the second book in this series *Science Beyond the Classroom Boundaries 7–11 Years* the approaches taken in this book are built upon and extended.

The list of skills provides a framework to stimulate a range of approaches directed at achieving a number of success criteria – for example, children will have:

- developed an understanding of what it means to be a good self-manager, communicator, problem solver, creator and teamworker
- tried out and modelled a range of strategies related to the skills
- reviewed with others their skills, taking and giving feedback
- become more articulate at describing, explaining and exemplifying how the skills have affected their learning.

What is the reality?

In the initial stages of the early years encouraging the development of children's independence is central to teaching and learning. In the classroom teachers and other adults support and work alongside children by explaining, demonstrating, modelling and scaffolding, encouraging the children to make their own decisions and to explore their own ideas. This movement towards being independent and autonomous is exciting as well as daunting for both adults and children since it constantly shifts personal boundaries. Sadly, however, as children move from those first stages in early years within the primary curriculum, there appears to be a shift in approach that narrows the children's opportunities for science 'Beyond the Classroom Boundaries' and for developing Personal Capabilities, thus reducing the opportunities for children to develop their abilities as independent learners.

You might want to ask yourself whether in your school children begin their primary career where they have access to outdoors and are able to follow their own interests, then move into the next year group where teaching and learning in science:

- is based mainly indoors
- offers children fewer opportunities to be the decision maker in their own learning
- focuses on subject content
- is not underpinned sufficiently by the development of skills and Personal Capabilities.

When this happens there is a clear discontinuity between elements of early years education in science. We would suggest that children's science education would be more successful if practice remained closer to that of the initial year in school.

What are the options?

In this section we will share a model which describes some options for us to help children respond to and enhance their Personal Capabilities when developing science. As you read this section consider how this relates to developing children's ability to work in science 'Beyond the Classroom Boundaries'.

Step 1: Be explicit

One of the most important aspects in the early years is to develop the children's concept of what it is to be a learner in science, how to behave, how to work, and so on. They should be encouraged to become familiar with a range of basic Personal Capabilities: for example, teamwork is about sharing out jobs, taking turns, fulfilling commitments, helping others and sharing ideas. This can only be achieved if teachers and other adults working with children are explicit and draw the children's attention to how they learn and to associated behaviours. Recognition of how we are going about things is the first key step.

In Table 3.1 based on material from *Smart Science* (Bianchi and Barnett, 2006) we have outlined what is valued and shown what it might look like in relation to children engaged in an activity to collect invertebrates from around the school.

Figure 3.1 Model of Personal Capability development

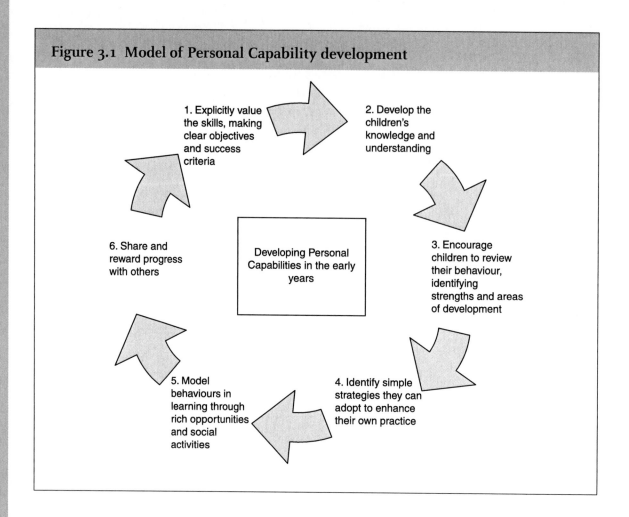

1. Explicitly value the skills, making clear objectives and success criteria

2. Develop the children's knowledge and understanding

3. Encourage children to review their behaviour, identifying strengths and areas of development

4. Identify simple strategies they can adopt to enhance their own practice

5. Model behaviours in learning through rich opportunities and social activities

6. Share and reward progress with others

Developing Personal Capabilities in the early years

Displaying these objectives on posters and boards outside in the school environment and sharing them with all members of the school community (including administrative staff, governors and parents) shows the children that these things are valued. Photographs, cartoons, logos, colours and clipart linked to keywords or symbols are ideal for getting key messages across, Encourage the children to find pictures or photographs that they feel represent capabilities such as good teamwork and good communication.

Step 2: Underpinning knowledge and understanding

Develop the children's knowledge and understanding of a skill or capability: help them make sense of what the skills mean to them. Explore what types of behaviour, people and activities they associate with different skills and capabilities (e.g. teamwork). Ask, Who in this school would you say works well in a team? Why? What does it mean to be a team player? You will be amazed at how much the children already know about such skills and possibly surprised about some misconceptions they may have.

Table 3.1 Personal Capabilities and working 'Beyond the Classroom Boundaries'

Activity: *Children are learning about invertebrates and have decided that they want to collect and observe invertebrates that are found in the school grounds. The teacher or adult working with the children provides quality intervention by asking a series of questions which lead children to reflect on what they are going to do. The adults use the language of Personal Capabilities as well as scientific words, such as invertebrates (as opposed to minibeasts), collect, observe and care for, return, habitat.*

Personal Capability	Examples to be able to:	Application 'Beyond the Classroom Boundaries' Collecting invertebrates
Self-management	Follow instructions Follow directions Stay on task	Sharing jobs Following simple instructions Meeting a deadline
Teamwork	Work with someone else Listen to other people's ideas Sharing jobs out and offering to take one on Taking turns	Deciding how to work. Agreeing on the need to stay together as a group. Making sure that each person has a job. Helping each other but not do someone else's job.
Creativity and problem solving	Talk about their ideas Be curious and ask questions about what they see	Asking questions linked to subject knowledge about invertebrates e.g. where they live?, what they need to live? Thinking about different objects and materials available and how to use them to make an invertebrate home. Solving problems if one approach does not work. Collecting and making a home for a snail, woodlouse, worm.
Communication	Share their ideas Listen to each other Respond to suggestions Talk about something that interests them	Telling other children about what they have done and what interests them.

33

As you begin to work 'Beyond the Classroom Boundaries' with children, it will be necessary to introduce them to key ideas and language related to Personal Capabilities. At the same time children will also need to develop their understanding of how Personal Capabilities will help them to work 'Beyond the Classroom Boundaries'. Some teachers found it useful to develop a range of scaffolding techniques to introduce and reinforce children's understanding of Personal Capabilities.

In this example we show how stories and simple games can be engaging, relevant and fun when introducing a Personal Capability or skill.

Using *Mr Men* Books

Some schools working with Personal Capabilities have been using cheap and cheerful books like the *Mr Men* and *Little Miss* series by Roger Hargreaves. These book can be used to engage children in discussion about the behaviour of the characters in the books and those that the children think should be valued.

Table 3.2 Using *Mr Men* and *Little Miss* for developing Personal Capabilities in science

Mr Men and *Little Miss* books	Supporting Personal Capabilities in science 'Beyond the Classroom Boundaries'
Little Miss Scatterbrain and *Mr Forgetful* Personal Management	This is a great story for helping children to think about managing themselves and helping them to realize that it is important to be organized, particularly when going outside to work. What would happen if Little Miss Scatterbrain forgot to take her science equipment outside when she was collecting invertebrates? What do you think she should do (and therefore the children) to make sure she was organized? For example, she could have made a list.
Mr Bossy Teamwork	Use *Mr Bossy* to introduce effective teamwork in science, ask children to think about what makes a good team, what would we tell Mr Bossy if we were explaining what children do if they are working well in a team, for example: • we have different jobs • we do our own jobs • we help each other • we listen to each other • we take turns

Mr Men and *Little Miss* books	Supporting Personal Capabilities in science 'Beyond the Classroom Boundaries'
Little Miss Chatterbox Communication	As we might expect Little Miss Chatterbox talks all the time and does not let anyone get a word in. This is a great context for challenging children to think about expectations for talking in science. Asking children to think about what Miss Chatterbox should really do will help to unpick the behaviours that you want children to value when they are engaged in discourse within their group or whole class. So Little Miss Chatterbox should think about: • not talking all of the time • counting to ten before she says anything • not to interrupt when someone else is talking • listen to and think about someone's ideas • think about whether she agrees or disagrees, or likes their ideas, before she says anything • when she talks she should think carefully about what she is saying and which words she is using • she should try to use science words when she talks All these statements can easily be converted to what the children should think about and do
Little Miss Shy Sharing ideas, being confident	One of the important elements of science is for children to feel confident and to be able to share their ideas about what they understand and how they think they might work. Some children can be afraid to share their ideas, perhaps uncertain of the response from others. Wherever children are working in science it is crucial that we can access what they are thinking, so that we can support them in moving forward in their learning. Using the *Little Miss Shy* story children can talk about how they could make sure everyone was involved and why this would be important. For example: • the shy person might have good ideas • the shy person might be able to help • the shy person might want to say something but be afraid of what other children might say From this discussion children could decide what they should do if they felt like that or to help someone else

(Continued Overleaf)

Table 3.2 Continued	
Mr Men and *Little Miss* books	Supporting Personal Capabilities in science 'Beyond the Classroom Boundaries'
Mr Impossible Tenacity, perseverance, problem solving	In science and particularly when children are working 'Beyond the Classroom Boundaries' we would expect that they work more independently and when faced with problems try to solve them. This often requires children to persist when they might give up and persevere to find a solution instead of immediately seeking help from an adult. The great thing about the character Mr Impossible is that he makes things possible. So try asking children how he does that – perhaps he: ● tries and does not give up ● thinks positively ● is not afraid to try something he has not done or thought about before Perseverance and the ability to problem solve are Personal Capabilities that are important to science and especially when working outdoors, and children need to know that they these are valued.

In taking this approach we are building up the success criteria on which children can base some reflective discussions. By developing understanding in this way children build a knowledge base relating to Personal Capabilities when working away from the classroom – in other words, what kind of behaviour they should expect of themselves and each other when they are working in science 'Beyond the Classroom Boundaries'.

Once children are familiar with the language behind a skill or capability, highlight one or two objectives for the task. Talk about these alongside the science objectives in your lesson: for instance, explain to the children that you will be looking to focus on the objective 'to take turns' while exploring living things in various habitats around the school. As such you will be providing them with double objectives: one focused on the process of learning (the Personal Capability) and one focused on the product (the science knowledge, understanding or skill).

Step 3: Self- and peer-review

Provide fun opportunities for children to think about their skills, what they are good at and what they think is good in others using self- and peer-review strategies.

There are many approaches, some as simple as verbal strategies, like 'Thumbs-up-sideways-or-down approach' (good, not sure, no) or traffic lights (green for something that went well, amber for something that could be improved, and red for when more help is needed) can be used when reviewing activities against the Personal Capability and science objective.

The drive for self- and peer-assessment has increased since the launch of Assessment for Learning in the UK. It is important that children working 'Beyond the Classroom Boundaries' are encouraged to think about how well they have worked, so that they celebrate and reinforce the development of Personal Capabilities.

Thumbs-up

Schools have taken the 'thumbs-up' approach a step further by helping children to think about their thinking (metacognition) by challenging them to think beyond the initial judgement. For example, if children give a thumbs-up for having been co-operative when working outside (1st stage metacognition) they are asked to describe how they did that (2nd stage metacognition) and then pushed to explain why doing those particular things was helpful (3rd stage metacognition). Using peers as reflective partners also helps objectivity, and supports the celebration of positive behaviours. If you would like to find out more about this you can view some video clips on the AstraZeneca Science Teaching Trust (AZSTT) site by clicking on the Personal Capabilities CPD Unit (www.azteachscience.co.uk/resources/cpd/personal-capabilities/view-online.aspx).

Peer-review

Children can be asked to nominate someone who they felt displayed a Personal Capability: for example, being organized in science by looking after the science equipment in their group when working outside, or persevering when they had a problem with trying to make a good bubble mixture.

Step 4: Simple strategies towards success

Encourage children to be more independent and organize themselves when they are going to work in science 'Beyond the Classroom Boundaries'. For example, if they decide to go on a 'Minibeast Safari', encourage them to stop, think and plan what they will do and what they will need. The Science Haversacks and Trolley described in Chapter 7 can really help children to think forward and support the development of Personal Capabilities such as self-management.

Infusing or embedding skills and capabilities requires a proactive approach. We need to think of ways in which we can guide and facilitate children's learning during a task, so that their activities encompass both the development of the skill and the science subject knowledge. It is not enough for us to expect children to know how to do the personal skills element of the activity, we need to help them to recognize the Personal Capabilities that they need to work on, such as identifying and taking on different roles in a group.

When working in the more flexible outdoors environment we must be even more proactive in providing the structures to help them out, for example, ways of organizing themselves, listening protocols to help each participant share their ideas, or time guides to assist them with managing a task to meet a deadline: for example, giving them a sand timer or an egg timer that pings when their time is up.

Step 5: Rich, embedded experiences – Outdoor Learning Partners

Teachers from project schools are beginning to adopt a range of different strategies when working 'Beyond the Classroom Boundaries' that includes asking the children to identify an Outdoor Learning Partner (OLP). Explain to the children that their OLP will be the person with whom they discuss the task, clarify what needs to be done, consider what they might need and come up with initial suggestions of how to do it. This could be a dedicated time for planning to help children, particularly older children, to think through what they will do, how they will record, who will do which job and what they need to collect to tell or show others. Some schools have started to extend the OLP from within a year group to pair younger children with an older child on a regular basis. This has many advantages, for example:

- when working around the school grounds an older child can take responsibility for working with children a little further away from the classroom than normal
- an older child can take the part of the 'expert': for example, in subject knowledge, or teaching a younger child how to use an unfamiliar piece of equipment
- while offering support to younger children, it also helps the older child to develop a range of Personal Capabilities and confidence in science.

Step 6: Rewarding progress

The use of whole-school- and classroom-based rewards are useful in recognizing and endorsing positive behaviours with regard to Personal Capability development.

Award systems

Many schools have adopted award systems to support the development of Personal Capabilities in science and when working 'Beyond the Classroom Boundaries'. Certificates of merit, Star of the Week Boards, stickers and Golden Time are just a very few of the many ideas schools regularly use to reward achievement, effort and learning of Personal Capabilities. Teachers encourage children to identify and explain who should be rewarded and why. As much as possible such rewards should celebrate effort and persistence towards developing oneself.

As with all learning it is important to be positive: Personal Capabilities are strengths to support and enhance self-esteem and self-image.

What will you do?

Life skills not classroom skills: get as many people in the school involved as possible, emphasizing the same types of skills and using the same type of language. The more children see that there are consistent messages about these behaviours in and around the school the better. Remember that these should be shared with parents so that parents appreciate that they are as important as other forms of learning such as reading and writing.

Take and allow some risks: challenge yourself to step outside the normal structured, teacher-led or highly scaffolded learning experiences, so that the children's skills and capabilities are challenged and stretched. This could relate to giving more choice or freedom with regard to who they learn with, what they learn and indeed where they learn! For example, when children are planning an investigation, include as part of that planning: who they are going to work with, what they need to do and where they think it will be best to work, making sure that they understand that they can choose to work in science 'Beyond the Classroom Boundaries'. Preparing for these risks will help you be more confident to take them, so talk about boundaries and strategies to show skills being used and of course sanctions where necessary.

Learner voice: Take learner views seriously; create opportunities for children to share stories and strategies, and seek ideas from other children or adults. It is essential that opinions are valued if children are to feel confident to engage with peer review. Enhancing these opportunities will develop their confidence and trust in themselves – that they are taking responsibility for their own development. For example, provide time for children to discuss their ideas about working outside with the rest of the class.

Active questioning: Develop an atmosphere where learners ask questions of you and others about the way they can work or have worked. Ask them to question what challenges them now and how they will manage those challenges.

Catch confidence: Proactively develop ways to celebrate effort and progress with regard to Personal Capabilities by using self-review and peer-feedback, as well as adult praise when they are successful.

Think smart, plan smart: Develop your own and children's sense of progression in Personal Capabilities, so that you and the children think about how within a few weeks, over a couple of terms or the school year, Personal Capabilities can be developed. Talk and plan as a staff how your school provision can be more responsive to children's personal development needs and requirements.

Practical ideas for teaching Personal Capabilities through science 'Beyond the Classroom Boundaries'

Making a shelter

Sarah at Castleside Primary School in the north-east of England worked with colleagues to develop Smart Science Capabilities across the school alongside universal access to the outdoors. Over time it became evident that children were beginning to take the lead in their own learning. This 'story' illustrates an occasion when children led the learning and the teacher felt confident to step back and observe.

In the area used by Key Stage 1 there were six bamboo poles, approximately 5 feet high. Until Sarah began the process of leading staff in a move to use the outdoors more in science, the poles had been ornamental. As part of continuing professional development, staff were challenged to use this element in the school grounds to develop thinking and working in science. Her colleagues were wonderfully creative and came up with all sorts of ideas; gradually they began to use the school grounds more frequently to support science activities and even used the poles for hanging objects from, such as shiny objects and light catchers. Staff had inadvertently modelled the idea that the poles could have a use, and this led to a surprising episode with a mixed Foundation and Year 1 group.

Despite it raining one day the children decided that they still wanted to go outside and with absolutely no intervention from their teacher they found a variety of covers in their classroom and took them outside to see which one would be the best to make a protective canopy. They stood under the different covers to test if they were waterproof and chose a blue plastic one that had been left over from their pirate ship role play area. Staff were amazed that although it was not actually planned by the teacher and had very little intervention from her, the children still managed to work out a scientific investigation. The only assistance they required once their final choice had been made was to ask an adult to tie the cover onto the bamboo poles to create their rainproof canopy.

This provided such a rich context for developing a range of Personal Capabilities, from independence and perseverance, to teamwork and communicating their ideas to the teacher – so that she could tie the cover to the poles.

Try this with your own class

You might like to provide children with a context for shelter building: for example, perhaps the class topic might be 'Explorers' or 'The Camping Holiday' where they have to design, make and test their own shelter. Obviously in terms of science, key concepts should focus on materials and their properties, such as waterproof, windproof, flexible, strong, insulation (keeping children cool, shaded or warm) and Personal Capabilities could centre around teamwork, perseverance, organizing themselves, communicating with each other and listening to each other's ideas.

Before beginning, ask the children to think about which of the Roger Hargreaves characters they think would be most useful e.g. Mr Impossible, Mr Clever, Little Miss Helpful.

This activity will require:

- *Time* for children to complete it, which can help them to develop Personal Capabilities related to timekeeping, perseverance (not to give up if things do not work) and leaving an activity and coming back to it at a different time.

- *Choice* of materials to choose from, which means discussing ideas, listening to each other and perhaps changing their ideas and compromising.

- *Subject knowledge* because their shelter has to meet certain standards, which will demand that children share understanding about properties of materials and possibly challenge each others ideas.

- *Reflection and evaluation* – once they have completed their shelter, children should be asked to test their shelter and provide feedback to other children and adults on how good their shelter was and evaluate shelters made by other groups. At the end everyone should be challenged to think about what they have learnt about materials and making shelters, as well as which Personal Capabilities they think they have used and improved.

References

Bianchi, L. (2002) *Teachers' Experiences of Teaching of Personal Capabilities through the Science Curriculum*. Sheffield: The Centre for Science Education.

Bianchi, L. and Barnett, R. (2006) *Smart Science; Activating Personal Capabilities*. Sheffield: Sheffield Hallam University (available online at www.smart.science.co.uk).

Lucas, B. and Claxton, G. (2009) *Wider Skills for Learning*. London: NESTA.

4

Managing children working in science 'Beyond the Classroom Boundaries'

What is the goal?

In this chapter we consider how the teacher can manage children working 'Beyond the Classroom Boundaries' in science and how children can learn to manage themselves.

Underpinning science 'Beyond the Classroom Boundaries' is the idea of developing children as independent learners. In the context of working outside, our goal is to create an atmosphere and range of situations where: 'Children can gradually move from regulation by others to self-regulation if appropriate frameworks and strategies for eliciting and acting upon their views are put in place' (Futurelab, 2008: 24).

Early years settings (3–5 year olds) are used to children being outdoors and have systems whereby there is always at least one adult outdoors observing or working with the children. The questions that arise within those settings therefore relate to the nature and quality of the science-based experiences and interventions provided and how children are supported to initiate their own learning in science.

Many of the teachers who have worked with us on this project to develop science 'Beyond the Classroom Boundaries' have recognized that the methods used by their Foundation Stage colleagues to manage children working outdoors need to be continued in the subsequent year groups. They raised a number of important questions, including:

1 How do we develop similar areas to Foundation Stage so that we can continue children's development in working independently across science outdoors?

2 How do we overcome the issue of the weather?

3 How can children be managed/supported in order to behave and work safely when learning outside of the classroom?

4 How can we communicate effectively with children working 'Beyond the Classroom Boundaries'?

5 How can we encourage children to organize themselves in science when going to work outdoors?

6 How can we help children to manage their learning: for example, organize their own resources, time, jobs and how to collaborate?

7 How can we help children to be creative in their thinking and ways of working when engaged in science outdoors?

8 How can we help children to keep track of their own learning in science while working outdoors?

What's the reality?

The reality of science in many schools is that as children move through the primary years there are fewer opportunities for children to work outside in science, and even fewer to do so independently. As part of our work with schools we asked teachers what stopped them from working with children on a regular basis in science 'Beyond the Classroom Boundaries'. Here are just a few of the comments we have collected from teachers; perhaps you recognize some of the issues as your own:

Can't always see children – are they on task?

Weather – we've got a boggy field!

Parents' complaints regarding safety and getting dirty.

The cleaners don't like it – muddy or wet feet coming in and out of classrooms.

Lack of equipment and resources e.g. clipboards.

It's stressful when the children were off task, difficult to reign back in a large space.

Not enough time.

Not sure about which parts of the science curriculum would work outside.

Need support from teaching assistants and we don't have it during science lessons.

Lack of adequate clothing for wet, cold and sunny weather.

Limited or no space to store equipment.

Behaviour – the children getting giddy.

In order to achieve the goal of children working as independent learners outside of the classroom in science, we must not be put off by constantly justifying why we think we cannot or should not take learning outdoors. The old saying 'Where there's a will there's a way' certainly applied to project schools. Their practice varied from schools where teachers were already moving towards universal access to science 'Beyond the Classroom Boundaries' across their school on a regular basis, to those where teachers were at the beginning of their journey. It soon became evident that once they had accepted that they could not hide behind excuses for going outside in science, then anything seemed possible and the prospect of changing practice became exciting.

When thinking about moving science 'Beyond the Classroom Boundaries' we need to think about two types of space: *our head space* – getting the children and teachers confident and in the right frame of mind; and the *physical space* – organizing areas or zones within the school that are safe places for children to think, to act and to learn in science.

Sorting out 'our head space' focuses on encouraging all staff in the school to have a shared understanding of the potential and purpose of moving 'Beyond the Classroom Boundaries' in science so that it is the norm not the exception.

What are the options?

In this section we share the suggestions from project teachers relating to the questions the teachers posed at the beginning of this chapter.

1 **How do we develop similar areas to the Foundation Stage so that we can continue children's development in working independently across science outdoors?**

In many schools the decision is influenced by the school building: for example, where the Foundation setting and Key Stage 1 classes were next to each other it was easy to develop a shared area so that the older children were able to continue to have outdoor experiences in science. The teachers then planned (often together) to ensure that the activities clearly focused on extending and challenging the children in relation to science and Personal Capabilities. Where this was not possible, some schools were able to create a defined area immediately outside the classroom. It was easy for those teachers in classrooms with a door which led directly onto the school grounds to make the most of that immediate access to the outdoors and develop an 'open door' policy similar to that found in early years settings. Some decided to have a fence built to create an enclosed area, while others created 'safe zones' by coning off particular spaces, where

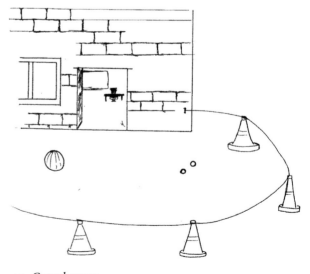

4.1 Coned areas

children were free to work, explore, pick things up, sit around and talk, think and work in science.

As children mature, not necessarily in age, but in responsibility, the safe zones should be modified accordingly. In other schools where the grounds were considered to be very safe staff did not want to have enclosed areas and decided that they would train children to work independently outdoors. In these contexts they created a set of working practices that everyone in the school understood, and children adhered to these when working outdoors in science and also across other curriculum areas.

A few schools had outdoor environments that were not easily accessible, usually because of the age of the building, so decisions centred around taking the whole class outdoors on a regular basis, ensuring that the science time outside was sufficient and used efficiently and with good outcomes for science and Personal Capabilities. When teachers reflected on how to manage taking a whole class from a second-floor classroom to the school grounds they realized that logistically it was no different to organizing an outdoor PE session.

2 What about the weather?

Having solved the issue of defining the working area, the next big issue for teachers was the weather. Indeed this was one of the first issues that prevented teachers from working outside in science.

> *Rain rain go away,*
> *Come again another day!*

Oh that it were so easy! Well actually it is. If early years settings (Foundation Stage) can get their head round the issue of weather with children who find it difficult to do up their own coats and shoelaces, then surely it should not pose a problem for the rest of us and for older children. It is all about asking the question, 'Will learning beyond the classroom boundaries elevate my teaching approach and the children's learning potential today?' If the answer comes back yes, then don the kagool, put on the wellies and off you go – simple as that. Interestingly teachers in the project schools admitted that often the only reason for not going beyond the classroom was because they felt that it was cold or a little too wet for their liking. Significantly, the children were not bothered! We are often the ones who limit the scope for children's learning and once we acknowledge this then ways of overcoming the weather become more obvious.

Involving parents

Staff at Shaw Primary decided that the answer to outdoor work in science was to explain to parents the reasons for using the outdoors for science. They decided to inform parents about the plans that meant all children, from those in Foundation Stage all the way through to Year 6 would be working outside in science whatever the weather. They asked parents to make sure that their children were sent to school with outdoor coats and suitable footwear.

Obviously staff decided to keep some spare items just in case, but it did mean that children would be suitably kitted out, and parents just needed regular reminders via letters home, newsletters and the school website.

Parents' outdoor science area notice-board

At Castleside Primary staff decided that it would be really useful to develop a notice-board to inform parents:

- about the new ways or working outdoors in science
- what the outdoors has to offer for science
- of the science topics and activities outdoors
- about projects outdoors that needed help: for example, clearing areas that were overgrown
- about sending children in coats, shoes and wellies so that their child can work outdoors.

Clearly, these approaches are to ensure that children can make their own decisions about when they need to go outside to do their science, and when we achieve this goal we should hear more comments like this one:

'I think next time we will need to put our coats on ...'

Maeve (child at Shaw Primary)

3 How can children be managed/supported in order to behave and work safely when learning outside of the classroom?

Behaviour was invariably one of the top three questions raised by project schools. Concerns were expressed relating to behaviour management and the worry that if children work outside then they are more likely to be less well behaved, whether the teacher was with them or not.

Teachers from project schools found it that it was a matter of setting some clear boundaries, establishing positive working practices and accepting that something might go a little off track in the early stages of working 'Beyond the Classroom Boundaries'.

Across the schools there was a realization that central to working in science 'Beyond the Classroom Boundaries' was the requirement to help children develop an understanding that they would be spending increasing amounts of time outside the classroom working in science and that there are expectations in relation to behaviour and ways of working. Their personal skills and capabilities were found to be fundamental to being in the right 'head space' when they worked outside. The implication is that children need to be encouraged to think about how they will behave, react and learn in this new outdoor environment, where they are more accustomed to being free to play. This will mean that the teacher will need to talk with children about working outside and the strategies that help them to be proactive, to think, to cope with challenge and to maintain self-control.

This may require extending classroom rules or establishing new ground rules for learning outside, reminding children about established school and classroom routines, and that these do not necessarily change just because they are outdoors. Children do require more than just a set of rules. They need to know how to work effectively in groups or teams, how to persevere with a different task, how to share what they have learnt, what they can do to sort out a problem before asking the teacher, how to keep track of and manage jobs and time and other vital learning skills. In other words, in order to be successful when they engage in science outside the confines of the classroom, children need to acquire the skills for managing their learning.

You might like to consider the following:

- engaging children in discussion about whether, when and how they need to work outside during a topic
- creating a rota which indicates which groups have access to the outdoors and when
- allowing a certain number of children outside for a set time; children are given special bands or bibs to wear, and when the set number of bibs has gone, no more children can go out
- training children to log out and log in using a white board, with a maximum number of children allowed out at any one time.

Here are some examples of how teachers from different schools approached this issue:

Start from where the children are at

Some teachers decided to start from where the teachers and children were at in different year groups. In Foundation Stage, adults and children alike were used to being outdoors where children work independently and with limited adult support. They engaged in a wide range of science-based activities such as making dens in the garden, choosing from a range of materials such as milk crates and tyres to plastic sheeting.

In subsequent early years classes teachers and children were less familiar with the daily access to working outdoors, and here staff took a softly, softly approach for both teachers and children. They began by arranging to take one group of children out at a time, supervised by an adult, while the rest of the class remained indoors working. This was a highly structured approach initially, using directed activities where it was easier to control the children's activity and behaviour. What it did was ensure success for teachers and children, which developed confidence and capabilities, and encouraged further use of the outdoors for science on a regular and less teacher-directed basis.

4 How can we communicate effectively with children working 'Beyond the Classroom Boundaries'?

As a class begins to move within the new learning parameters 'Beyond the Classroom Boundaries' both teachers and children might need the security of keeping in contact with each other. Project schools found interesting and imaginative ways of doing this, for example:

- *Walkie talkies*: these were chosen by some teachers who decided that the children would really enjoy using them, and that where children were working independently and out of sight of the teacher this would be an excellent way of communicating and also support the extension and review of learning during tasks.

- *Science Bibs*: these have been very popular with project schools; some used PE Bibs, while others purchased special Science Bibs with planets on them. This provides a visual communication which tells any adult in the school that the children wearing these are carrying out science tasks 'Beyond the Classroom Boundaries'. Interestingly children are proud to wear these, and also realized that they could be challenged by any member of staff if they were not working appropriately.

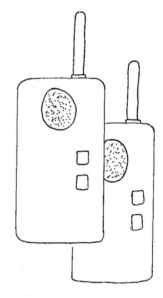

4.2 *Walkie talkies*

Equally, schools using this approach also agreed that staff should compliment children seen working well outdoors. The Bibs help everyone feel secure that children are not just wandering around the school and as a relatively cheap and cheerful resource these transmit key messages of:

- 'I'm learning', 'My teacher is aware I'm out of the classroom.'

- 'Please do not disturb – job in progress.'

- 'Do come and ask and we will explain what we are doing.'

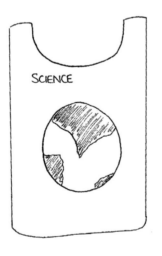

5 How can we encourage children to organize themselves in science outdoors e.g. time and roles?

4.3 *Science Bibs*

Scaffolding children to manage their own learning in science 'Beyond the Classroom Boundaries' can begin with something as simple as putting children into 'role' through using props such as 'Science Haversacks' (see Chapter 7). Here children know that they are going to work outside because they have been given a Science Haversack in which they place the equipment they need before they go out. This is the first step in helping children take the role of scientist, to choose what they need and to organize their equipment so that they are ready to go outside.

A number of teachers found this a very useful approach: noting that children began to think differently about going outside, and realizing that they needed to be prepared for when they were going outside the classroom to work in science.

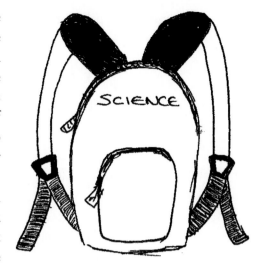

4.4 Science Haversack

Teachers also began to consider how to use resources such as the Science Haversacks to support other Personal Capabilities.

Time management: Teachers extended the use of the Science Haversack by tailoring it to supporting and developing a range of Personal Capabilities such as time management by including timers which beep when the children are due to return to the classroom or allocated place.

Roles: *Badges* were suggested to help children to organize themselves: for example, if children have decided to go outside to find objects made from different materials then they could wear badges and take on the following roles:

4.5 Personal Capabilities Role Badges

- *Timekeeper*: How long have we got, how much time do we have left?

- *Recorder*: How will we record which materials we find?

- *Communicator*: How is it best to tell the rest of the class what we have found out?

- *Task manager*: What have we got to do and who will tackle what? Have we done it well enough?

6 How can we help children to be creative in their thinking and ways of working when engaged in science outdoors?

We found that across project schools where teachers began to use the outdoors across the science curriculum their practice changed and teachers became more creative in their approach and more willing to take well-planned risks in the way they engaged children in science. Here are just a few examples.

Chalk bodies

In an early years class at Castleside the teacher changed the usual indoors activity on the body, where children drew around each other on paper, to children drawing around each other using chalk on the playground. The children were then challenged to use anything safe from the grounds to measure themselves. One child chose to use twigs that were on the ground underneath the trees and another child used stones. Going outdoors changes the potential outcomes, leads to creative suggestions from teachers and creative responses from the children.

Testing coats

Normally, Sarah from Castleside made mini-coats for the children from different materials and they tested them indoors to see if they were waterproof, using pieces of fabrics secured over a plastic containers and pipettes to drip water onto the fabric. You are undoubtedly familiar with this activity; it is quite teacher-directed with little room for creative input from the children.

However, once Sarah started to think about moving many of her science activities 'Beyond the Classroom Boundaries' she found that she became more creative in her response and the children rose to the challenges that she set.

You might like to think about how taking activities outside changes the way that children respond to the problems: for example, as Sarah did by taking the popular Teddy's Raincoat activity outside without the usual prompts of fabric, containers and pipettes so see how the children would respond. Freed from these constraints children are more likely to think about using the environment to test different materials because it provides children with a totally different context, one where they can think about:

- using mud to make the coats messy and washing them to see how easily dirt comes out of the fabrics

- using watering cans to pour water over the fabrics to find out if they were waterproof and a washing line to see how quickly different fabrics dry

- rubbing fabrics on stones to simulate children playing on rough surfaces

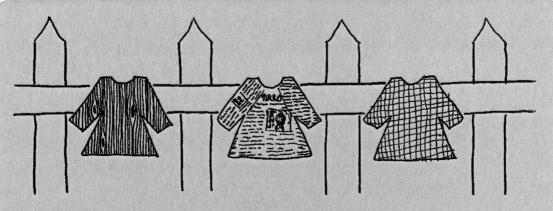

4.6 Teddy's Raincoat

Sarah and her class actually decided to create mini-raincoats of different materials, hang them from the railings and leave them there for several weeks to see what happened to the different fabrics. This was a great idea and one which also introduced children to the experience of carrying out an investigation over a longer period of time, where they had to take regular observations to record changes in the coats.

8 How can we help children to think about their own learning in science while working outdoors?

Some of the skills children need to manage their work outdoors effectively can be addressed in this way; however, the affective or motivational aspects of learning during a task are more difficult to scaffold as they rely on personal commitment to learning. Whether a child avoids giving up easily or takes time to be imaginative are aspects of learning that generally only come about when these are seen as valued behaviours. For these capabilities visual cues, posters, recognition and rewards are initial steps as a reminder to children of the worthiness of a particular capability and its value in learning. As such, badges or visual cues are all ways of making real and explicit the skills that lay the ground for successful working beyond the classroom.

You might like to consider the idea of Thinking Posts around the school to encourage children to think about their Personal Capabilities. These remind children about capabilities that are valued and also celebrate them, for example:

- 'When we work outside we help each other.'
- 'When we work outside we try hard to solve our own problems.'
- 'When we work outside we don't give up, we persevere.'

4.7 Thinking Posts

Shaw School is planning a set of Tibetan Science Flags which will be made by children in partnership with an artist in residence. Each flag will feature an aspect of science, and one of the flags will dedicated to Personal Capabilities and have printed on it words such as persevere, problem solve, help each other, think, be determined and independent. Carol Sampey's aim is to recreate the approaches used in school, where the indoor environment contains prompts to remind children of ways to work. She also wants to make very public to children, staff, parents and visitors the attitudes and personal competencies that are valued, related to working in science outside.

4.8 Shaw School Tibetan Science Flags

Children being 'in charge' of their own learning is commendable, but only if we are sure that the learning is authentic, on track and purposeful. During the research for this book we found that teachers quickly identified areas of the curriculum and activities that could move outdoors. Project teachers found the issues they faced indoors were the same when children were working outdoors. Indoors, teachers needed to check that children understood the activity and concept, and the same was required when children were working 'Beyond the Classroom Boundaries'. Project teachers developed a range of responses to this issue: for example, by giving children a set point in the activity to return to the teacher to check that they are on task.

Green flag system

Kay at Wheatlands Primary in Redcar adapted the flag system from Sheffield schools initially with older primary children and eventually with younger year groups. She placed one of these flags in the classroom window to which the children had to respond.

- *Red flag* means come back into the classroom immediately.
- *Amber flag* means that you have 5 minutes and then the red flag will be in the window.
- *Green flag* means that you can keep working outdoors.

One of the things that Kay is trying to establish at Wheatlands Primary is a set of working practices that begin with the early years classes and continue throughout the school, so that the children 'grow up' with a set of rules that are consistent and that become second nature to children and to staff.

Celebrate Personal Capabilities

As with all learning and that includes children working on their own Personal Capabilities, it is important to value and celebrate the children's achievements. So when children have completed their tasks 'Beyond the Classroom Boundaries' do encourage the children to talk about what they did and found out in relation to science and also about how they worked, which Personal Capabilities they used and those that they think still needed some work. You might give children stickers, points or certificates of achievement which not only celebrate children's accomplishments but also help children to understand what kind of thinking and behaviours are valued when working in science 'Beyond the Classroom Boundaries'.

What will you do?

Project schools found that the best way to work was as a whole staff, so that eventually a whole-school way of working was developed, as Kay at Wheatlands Primary was trying to establish in her school. The starting point was invariably to provide a forum, usually at a twilight staff meeting or inset day, where staff could raise concerns and issues about working 'Beyond the Classroom Boundaries' in science. Once these were acknowledged the next stage was to challenge staff to come up with positive ways to overcome those concerns.

Remember that it is as important to prepare the children as it is for you to have planned the spaces and experiences for them. Essential to this is a clear understanding of your expectations for learning outdoors. Share your thinking with children and model positive behaviours and practices; start with shorter focused activities and work towards more open-ended, team-based challenges.

Practical ideas for teaching Space 'Beyond the Classroom Boundaries'

Have you ever taken the topic 'Space' into the space outdoors? It can transform this project. Here are just some ideas of how to do this, have a go with your class or, indeed, with all of the early years classes in your school.

An alien landed in our school grounds

How about this engaging idea of creating a very imaginative scene in your school grounds? Create the equivalent of a spaceship crash on the grassed area and when the children in early years classes arrive in school they find the area cordoned off with police tape (do warn parents what you are going to do, so they do not think it is

4.9 *Alien Spaceship Crash Scene*

for real!). Make sure that staff pretend to be equally surprised and go along with the children, support them in looking at the evidence and thinking about:

- What was it?
- Where did it come from?
- How did it get there?
- What were all of the different pieces for?
- How could they get a message to the 'aliens' to say their spaceship was in the school grounds?
- Which planet did they think it came from?

In small groups of twos and threes allow the children to explore the crash scene, which of course is a defined area of the school grounds. It could include the use of some fairly imaginative items including parts from an old washing machine together with the drum and hoses!

This lends itself to children:

- taking the role of forensic detectives in looking for and thinking about evidence
- using chalks on the playground surface to draw what they thought the spaceship looked like before it crashed
- sharing ideas and working as a team to make decisions about when it landed, what it looked like and what the different parts of the craft were for.

Taking this approach can then lead to teachers and children thinking about:

- What was it like in space?
- Could we build our own spaceship?
- What materials could we use?
- Could we make out own spacesuit?
- What did the astronauts wear? What materials were needed?
- Where would we travel to? Which planet? What are the planets in our solar system like? Which planet could we land on?
- What is our solar system like? (Here children could use books and small groups could be in charge of drawing a planet each on the school playground.)

Many of the science resource catalogues have rockets that the children can send into the air, as the examples in Illustration 4.10 show.

Children can explore making these rockets fire into the air, what kind of forces make the rocket move, why it returns to earth, how to make the rocket travel higher and further across the playground or field.

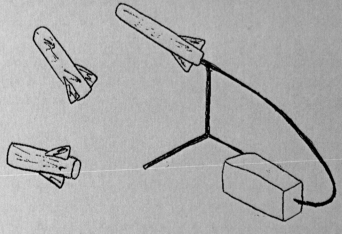

4.10 Mini rockets

Of course, this topic has to have children building their own spaceship either in small groups or the whole class creating a huge spaceship for children to use in role-play scenarios. Make it as close to reality as you can, making sure that the children wear spacesuits to enter, have breathing apparatus on their backs (children love to design and make these using plastic bottles), have portals which have transparent plastic, and get the children to take space rations: for example, dried food such as dried fruit and biscuits.

Children could watch video clips of space walks and astronauts on the outside of the spaceship fixing broken equipment, to see how they move, and add this to their role play. If we are really going to inspire children, how about giving them a basic video or Flip Camera and ask them to recreate the space walk. Encourage them to talk about what they are doing and how they are feeling.

In fact, the scope of this is only limited by your own and the children's imaginations. However, remember at the end of the day we hope to develop children's scientific concepts and skills as well as Personal Capabilities. So do make sure that you focus on the following ideas:

- pushes and pulls as forces that make things move, change shape and direction and stop
- gravity is a force that pulls things down to Earth
- we need to use special materials for spaceships and spacesuits
- in space it is different than on Earth; we cannot breathe in space and gravity is not the same
- the Earth is a planet
- we are part of the Solar System
- in our Solar System there is the Sun, other planets and moons.

Reference

Futurelab (2008) *Reimagining Outdoor Learning Spaces: Primary Capital, Co-design and Educational Transformation* (A Futurelab Handbook). Bristol: Futurelab.

5 Health, safety and risk 'Beyond the Classroom Boundaries'

What is the goal?

All schools within the UK are required to attend to safeguarding issues to ensure that there are effective safeguarding systems and frameworks in place, and of course this relates to children working outdoors in the school grounds. Every school is unique in terms of what is 'Beyond the Classroom Boundaries' in their school grounds and it would be inappropriate to create a definitive approach to health, safety and risk. Therefore, in this chapter we aim to offer advice concerning working 'Beyond the Classroom Boundaries' and focus on the following:

- the responsibility of adults working with children
- developing children's Personal Capabilities to enable them to work safely outdoors
- whole school strategies for working safely outdoors.

If you and your school decide to be proactive in developing science 'Beyond the Classroom Boundaries' this should include a review of the school health and safety policy to ensure that it has a section relating to children working outdoors. Local Authorities such as the Norfolk County Council suggest that, when reviewing and possibly rewriting your outdoors policy, you should aim to include ideas from the following groups:

- the children
- all staff
- parents
- governors and managers
- the wider community (the Local Authority, the garden centre).

We advocate that you take time to ensure that the process is thorough and the following questions could provide a framework for review:

- Where are we now?
- Where do we want to go?

- How can we get there?
- How will we monitor success?

The goal should be to have a policy where everyone, including the children, is a partner and shares responsibility in understanding and managing the health, safety and risk inherent to learning 'Beyond the Classroom Boundaries'.

What is the reality?

While it is everyone's duty to ensure safe working practices, 'Beyond the Classroom Boundaries' requires a hierarchy of roles and responsibilities in relation to health and safety when children work in the school grounds.

Table 5.1 Roles and responsibilities in relation to health and safety 'Beyond the Classroom Boundaries'

Governors	Ensure that documentation is in place and that it covers working outdoors, and it is monitored and reviewed regularly.

Head teacher	Has a duty of care towards everyone working outdoors.
	    Implements elements of the policy on a day-to-day basis linked to working outdoors.
	    Ensures a positive culture towards health and safety rules relating to outdoors and that they are followed by everyone.

Science leader	Ensures that the staff are aware of relevant health and safety issues in science.
	    Carries out risk assessments regarding the school grounds and science equipment.
	    Inform staff of where science safety publications are kept.
	    Support staff in developing and carrying whole-school health and safety policies in science.

Staff (including teaching assistants)	Be familiar, and comply with health and safety arrangements for working outdoors.
	    Report any problems or incidents immediately.
	    Indicate potential risks and how they will be managed in lesson planning.
	    Consult science leader when in doubt.
	    Set a good example to others.
	    Develop safe practices with children and their responsibility to manage risk.

Children	Self-manage risk to make sure that they take responsibility for their own safety and the safety of others.

All schools should have a copy of the Association for Science Education (ASE) (2011) publication *Be Safe!*; preferably more than one, so that all staff have their own copy or know where to access one. Many local authorities are members of CLEAPSS® and this means that primary schools in those authorities can have access to the website and extensive materials relating to health and safety in primary science, and in particular to working outdoors. Both ASE and CLEAPSS® welcome telephone or email queries from teachers regarding health and safety in schools. Information about both of these organizations can be found at the end of this chapter.

Health and safety in primary science is generally a matter of common sense, with some exceptions relating to specific activities or equipment that are covered in the safety documentation mentioned previously. The use of school grounds should be covered in school documentation relating to Safeguarding Children which would include all activities where the school grounds are used, such as playtimes and outdoor PE sessions. Hazards will have been identified and action taken where necessary. The school will already have in place rules and procedures: for example, children should not climb trees in the school grounds or children should stay away from the car park area, and this would apply for whatever reason the children are outside.

What are the options?

When children are engaged in science outdoors the risk factors change very little from when they are outside at playtime, when engaged in a PE lesson or moving around the school generally. Therefore, there are very few new health and safety issues relating to children working 'Beyond the Classroom Boundaries' in science that staff and children have to consider.

Once you have made sure that health and safety issues have been managed and actioned, through the general school policy, attention can then turn towards how to support the children to work outdoors successfully and safely. There are a number of issues that should concern us in science and they are listed below.

1 Children's understanding of the concept of a hazard and of risk.
2 The use of specific equipment that might pose a health and safety issue (e.g. binoculars (children should not look directly at the Sun)).
3 Children collecting or handling plants and animals.
4 The development of children's Personal Capabilities in science to manage themselves and others safely when working 'Beyond the Classroom Boundaries'.

The first crucial point in developing children's understanding of safety and dovetails with their personal development of self-management skills is their understanding of the concepts of hazard and risk. There is a difference between hazard and risk, and this can be explained to children in simple terms in the context of their science. A hazard is something that can cause harm: for example, eating potentially poisonous berries from a tree, or looking directly at the Sun. A risk, on the other hand, is the chance that a hazard will actually cause harm. For example, if there were no trees with berries then there would be no risk of being poisoned; if you never look directly at the Sun then the risk of damaging your eyesight in this way does not exist. However, if you occasionally look directly at the Sun your risk increases, and if you stared at the Sun for a long time then the risk of damage to your eyes is very high.

So, when we consider the second point, most things in life could be hazardous, and that includes using primary science equipment. However, the risks are limited if we act responsibly and use equipment safely, but increased if we misuse our equipment. When we consider the third point, plants and animals in the school grounds can pose hazards since some plants are poisonous, and climbing a tree can be dangerous. However, school policies should have covered these with statements such as:

- the caretaker/gardener checks regularly to ensure that no poisonous plants are growing in the school grounds
- low branches are cut so that children cannot easily climb trees in the school grounds.

Potential hazards are therefore limited and the risk of poisoning or children falling from a height is minimized. It is then up to the teacher and children, and indeed it is beneficial for children's education to follow some basic safety rules when using equipment and working outdoors, such as:

- if you are using sound makers do not make loud sounds next to someone's ear because their hearing could be damaged
- wash your hands after you have been collecting invertebrates.

Most national curricula for primary science demand that children develop their understanding of safe ways of working, adopt safe practices and manage risk in their science activities. We would expect children to progress through the primary years, with the teacher helping them to recognize risks and eventually for children to actively control risks to themselves and others, and this should begin in the early years.

When children work inside the classroom we expect them to work safely, so we should expect no less of them when they work beyond the classroom. The best way forward is to work with children as partners in understanding potential hazards and in minimizing risk to themselves and others.

Some schools have developed a whole-school approach to sharing responsibility with pupils. At Castleside Primary School each class created its own set of rules for working 'Beyond the Classroom Boundaries'. This was to ensure that all children were involved and that they worked at their personal ability level, as well as taking ownership of this issue. Interestingly, all the sets of rules had common statements, so by default there was parity in approach across the year groups, which gave the sets of rules a whole school feel. Once the rules had been agreed the expectation was that all children would:

- abide by the rules
- work together to make sure that whole groups kept to the rules
- take their roles seriously and would confidently challenge a classmates if their behaviour put them or others at risk.

At Shaw, staff were confident in children working 'Beyond the Classroom Boundaries' if the children were in sight of an adult. However, they still considered the issue of the safety of children when they were working independently, out of sight. So they decided that, rather than allowing children to work on their own or in pairs, they would take another approach.

Health and safety at Shaw Primary

At Shaw, in the Foundation Stage, children worked in an enclosed area, so the issue was developing children's ability to use science equipment appropriately. Outside this area children were closely supervised but encouraged to talk through how to stay safe when working in the wider school grounds.

In Key Stage 1 children were able to work directly outside the classroom and elsewhere in the grounds which were deemed as very secure. However, there were sets of rules for working away from the classroom which the children understood and which demanded that they behaved responsibly keeping themselves and their friends safe. This of course was no different from any other element of school life: if they worked away from the view of an adult they were usually in groups of at least three children, and if, despite all the strategies, something happened the children had an understanding of the protocol for getting help – for example, one child stay with the injured pupil and another goes to get help from a teacher or other adult.

This link to Personal Capabilities, specifically teamwork, encourages children to consider and manage risks to themselves and to others in their group. It also draws on communication, ensuring that children are confident to communicate concerns about safety to each other. Thus children are engaged in self-management because they have to ensure that they take responsibility for working safely.

Developing guidelines with children prompts positive action because it allows the children greater control of their own performance.

An alternative approach was developed by Kay at Wheatlands Primary School and is being adopted by staff in most classes.

Red, amber and green flags

Kay tells of how she worked with a colleague who was a Year 4 teacher: 'We developed a flag system using laminated coloured sheets of card. When the children were using the outdoor space independently they knew their parameters, that they weren't allowed to go further than X, Y and Z. If they saw the green flag in the window then knew they were free to continue their task. If they saw the amber sign come into the window they knew they had approximately 5 minutes to finish off. Then if the red flag went up in the window they had to come back immediately. The children were really good because every now and then I would test them and put a red one up and they would all come back, it worked really, really well. My year 4 colleague commented that it was really good how the children were mindful of what they were doing yet also kept an eye out for the flags as well. She was surprised at how quickly they responded when the red flag went up. We are both going to use it with future classes, already other staff are taking this approach on board, having seen it work with us, so I am sure that it will filter through the rest of the school, including the early years classes.

What we should not do is underestimate children's ability to understand and respond to working safely outdoors. Of course, it becomes easier if it is part of our everyday discussions in science with children. What we cannot afford to do is to sanitize science, while science curricula demand children understand safety, common sense should tell us that children of all ages must be involved in making decisions about how they and those children they are working with can stay safe. Alice (age 6) understood this when she decided to work outside:

> We did hearts outside, when you haven't done exercises it did not beat much, but when we have done jogging or running it goes faster. Your heart is like a pump round your body. It is better do it outside, you might break something inside.

Alternative perceptions of 'risk'

When we discuss risk in primary science, whether it is when children are working inside or outside the classroom we usually think about health and safety issues. However, risk can relate to different aspects of out lives: physical, mental and emotional. Risk is relative to the situation and the capabilities of the individual to deal with a situation. There are many different types of risk in primary science – for example, risk can mean:

- sharing your ideas with other people because you are not sure what their reaction might be
- trying a different approach because you have never done it before
- trying something new, such as using a pooter, because you are not sure if you are using it correctly and might swallow an insect

- overcoming physical obstacles relating to walking on uneven surfaces, balance or manipulative skills

- using equipment safely, such as binoculars and not looking directly at the Sun or making sure that you lay down on the ground by the pond when dipping for pond animals so that you cannot fall in the water

- being allowed to go outside for the first time with a science partner and to work away from the classroom without direct supervision of the teacher.

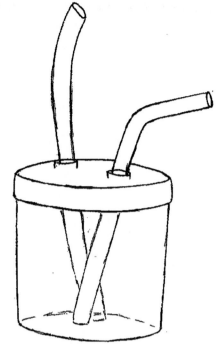

5.1 Pooter

Project schools raised an important issue about the difference between science inside the classroom and science 'Beyond the Classroom Boundaries'; they were concerned with the distance it creates between pupils and teacher. However, we can view the distance between teacher and children positively rather than a worrying situation. That small distance can help to develop children's independence and management of personal risk in different ways: emotional, physical and mental. In the classroom the teacher might be oversupportive because it only takes a few steps to get to a child. Whereas when working in the school grounds it may take a little longer to reach children, which means that the teacher is more likely to wait and observe, decide who to go to and why. The children realize that attention is unlikely to be immediate and are more likely to take responsibility and to solve the problem themselves.

What will you do?

If we go back to the beginning of this chapter then the answer to 'What will you do?' must be to ask and answer the following set of questions:

- Where are we now?

- Where do we want to be?

- How can we get there?

- How will we monitor success?

The most successful approaches have been where staff and children across the school have been involved. The next section suggests some practical ways to involve children in using and developing their Personal Capabilities to work independently and safely 'Beyond the Classroom Boundaries'.

Practical ideas for developing children's ability to work safely and independently 'Beyond the Classroom Boundaries'

Understanding hazards and risks

As you begin to introduce children and indeed staff to working 'Beyond the Classroom Boundaries', the first issue is an understanding of the concepts of what is a hazard and what is a risk.

Think–pair–share

Ask children to discuss with their science talk partner about how they could make sure that when they work outside they stay safe.

Make a foursome

When children are happy with their ideas ask them to join with another pair, to share their thinking with another pair. This will certainly draw upon a range of Personal Capabilities such as listening to each other, turn taking, compromising and respecting ideas from others in their group.

Share with the class

This is the point where groups share their ideas with the rest of the class; and it is here that the teacher can help the class to recognize similarities in their ideas and help children to come towards agreed class ideas about keeping safe. Of course, if for example they suggest that 'a lion might get them', do challenge their understanding of what it means to be safe in the school grounds!

Children carrying out a safety check of the outdoors area

Why not take children outside to do a safety check of the school grounds. Give children large hazard signs and tell them that they are going to be safety inspectors, and they have to put a hazard sign anywhere that they see something that could be dangerous.

Children could also:

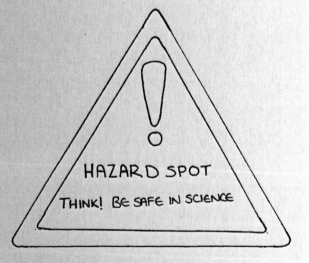

- be given a camera to take photographs of the danger
- be asked to create sentences to go with their photographs, explaining the danger and suggesting what children should do to keep safe

5.2 *Hazard spot*

Figure 5.1 Children's Risk Assessment Card

OUR SCIENCE RISK ASSESSMENT CARD

BE
SAFE!

Our activity

HAZARD	Why is it a hazard?	How do we keep ourselves safe?

Signed_____.

Date _____

- print their work out as danger signs and place them in the correct places around the school grounds.

Risk Assessment Cards/Forms

One of the key aims of this book is to encourage children to become independent learners who take charge of their science. Thus, when children work 'Beyond the Classroom Boundaries' they know that they need to stop and think about safe working practices. Then if they think that what they are going to do outdoors might pose some risk, the children make the decision to complete a Risk Assessment Card, and then hand it to the teacher or adult they are working with to 'sign off'.

Of course, such a system needs to be scaffolded, where the teacher helps to set up this way

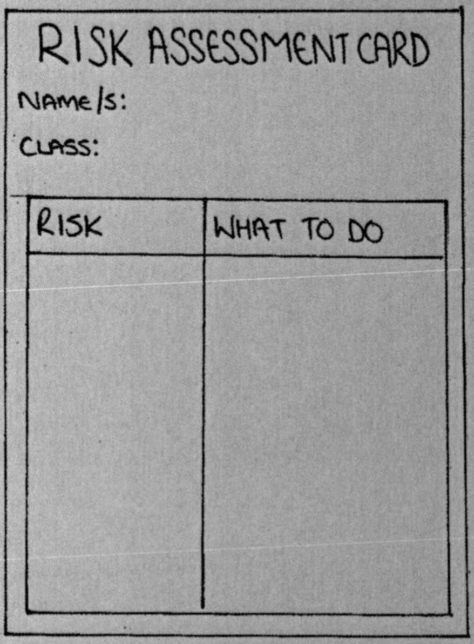

RISK ASSESSMENT CARD

NAME/S:

CLASS:

RISK	WHAT TO DO

5.3 Risk Assessment Card

of working and with adult support in the initial stages, children learn to use the cards or you may wish to create you own. Gradually as their personal understanding in science develops, and their appreciation of hazards and risk increases they become more capable of making their own decisions.

Health and Safety Monitors

In an earlier section of this chapter we indicated that it is the school governors and the head teacher who have the ultimate responsibility for safeguarding against issues around the school. We also suggested that on a daily basis everyone should contribute to health and safety, so it is appropriate that a school might think about appointing 'Health and Safety Monitors' who regularly walk around the school grounds and spot potential hazards. This is an excellent way to use and develop Personal Capabilities such as teamwork, responsibility, communication and motivation. For example, children could report a broken fence, glass or litter thrown over into the school grounds, loose stones that someone might trip over. The emphasis is on children identifying and reporting risk and hazard, which the school then attends to. In appointing 'Health and Safety Monitors' it is appropriate to give them badges so that everyone knows that they can report perceived risks, recognizable badges also promote the children's self-esteem. A lovely idea is for some children from Key Stage 1 to be given a 'Health and Safety Buddy' from Key Stage 2 and together perhaps once a week, they meet up and check the school grounds. When the Key Stage 1 child moves into Key Stage 2, he/she could then be a buddy for a children from early years.

References

ASE (Association for Science Education) (2011) *Be Safe!* (4th edn.). Hatfield: ASE.
Norfolk County Council Early Years Outdoor Learning (available online at www.schools. norfolk.gov.uk/myportal/custom/files_uploaded/uploaded_resources/5381/earlyyears toolkitfinalversion.pdf).

Contacts

Association for Science Education (ASE)
College Lane
Hatfield AL10 9AA

Tel: 01707 283000   email info@ase.org.uk

CLEAPSS®
The Gardiner Building
Brunel Science Park
Kingston Lane Uxbridge UB8 3PQ

Tel: +44 (0)1895 251496   email: science@cleapss.org.uk

CHAPTER 6

Progression and a review of your own progress

What is the goal?

Now that you have been introduced to the ideas related to working in science 'Beyond the Classroom Boundaries' it seems appropriate to suggest a pause to think and to consider the steps that could be taken towards competency and confidence in using the outdoors. In this chapter we introduce those progressive steps, known as the 'Escalator' (see Figures 6.1 and 6.2). Of course, movement does not happen overnight: in the Escalator we present some thoughts around what it could potentially 'look like' and 'feel like' for teachers and learners to move from practice which does not exploit working outdoors in science, to one where it is a regular and routine part of classroom life.

This chapter is designed to stimulate self-reflection and reflection on children's abilities and skills. You could use the Escalator on your own or with other teachers to review your current position and aspirations and to stimulate discussion with colleagues, governors, parents and indeed the children if you like, and as importantly as noting where you are and where you want to be, to objectively consider what it will take to move progressively up the Escalator. Throughout the chapter you will be referenced back to key sections of the book where approaches and resources are explained more fully.

What's the reality?

What works well for one person will invariably not be quite the same for the next. Progress is determined by where you start and is judged by where you hoped you would get to. In order to gauge where you are in your thinking and practice at the moment, we offer you two types of 'Escalator' – an 'Escalator of children's progress in using the outdoors for learning' (Fig. 6.1) and an 'Escalator of outdoor learning entitlement for teachers in schools' (Fig. 6.2). The escalators offer a six-stage model from what could be teachers or pupils who currently do not use the outdoors to others who use it as an integral part of teaching and learning and to those who have fully grasped and exploited its potential. Each stage is illustrated by a range of descriptions that aim to support reflection and assist us in identifying not only where we are but what we do particularly well at and what we might aspire to do more of.

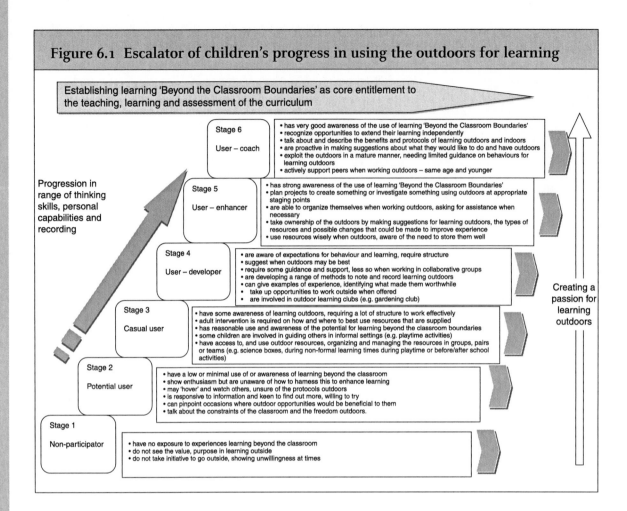

Figure 6.1 Escalator of children's progress in using the outdoors for learning

Establishing learning 'Beyond the Classroom Boundaries' as core entitlement to the teaching, learning and assessment of the curriculum

Progression in range of thinking skills, personal capabilities and recording

Stage 6

User – coach
- has very good awareness of the use of learning 'Beyond the Classroom Boundaries'
- recognize opportunities to extend their learning independently
- talk about and describe the benefits and protocols of learning outdoors and indoors
- are proactive in making suggestions about what they would like to do and have outdoors
- exploit the outdoors in a mature manner, needing limited guidance on behaviours for learning outdoors
- actively support peers when working outdoors – same age and younger

Stage 5

User – enhancer
- has strong awareness of the use of learning 'Beyond the Classroom Boundaries'
- plan projects to create something or investigate something using outdoors at appropriate staging points
- are able to organize themselves when working outdoors, asking for assistance when necessary
- take ownership of the outdoors by making suggestions for learning outdoors, the types of resources and possible changes that could be made to improve experience
- use resources wisely when outdoors, aware of the need to store them well

Stage 4

User – developer
- are aware of expectations for behaviour and learning, require structure
- suggest when outdoors may be best
- require some guidance and support, less so when working in collaborative groups
- are developing a range of methods to note and record learning outdoors
- can give examples of experience, identifying what made them worthwhile
- take up opportunities to work outside when offered
- are involved in outdoor learning clubs (e.g. gardening club)

Stage 3

Casual user
- have some awareness of learning outdoors, requiring a lot of structure to work effectively
- adult intervention is required on how and where to best use resources that are supplied
- has reasonable use and awareness of the potential for learning beyond the classroom boundaries
- some children are involved in guiding others in informal settings (e.g. playtime activities)
- have access to, and use outdoor resources, organizing and managing the resources in groups, pairs or teams (e.g. science boxes, during non-formal learning times during playtime or before/after school activities)

Stage 2

Potential user
- have a low or minimal use of or awareness of learning beyond the classroom
- show enthusiasm but are unaware of how to harness this to enhance learning
- may 'hover' and watch others, unsure of the protocols outdoors
- is responsive to information and keen to find out more, willing to try
- can pinpoint occasions where outdoor opportunities would be beneficial to them
- talk about the constraints of the classroom and the freedom outdoors.

Stage 1

Non-participator
- have no exposure to experiences learning beyond the classroom
- do not see the value, purpose in learning outside
- do not take initiative to go outside, showing unwillingness at times

Creating a passion for learning outdoors

So in order to consider your 'reality', take some time to look at each of the Escalators. The first considers your children and their competency at working outdoors and the second your position as a teacher and the provision you offer. Try to be as honest in your appraisal of yourself and in appraising your pupils. Do not make excuses or look for reasons to justify your position, just make as honest a judgement as you can using a 'best fit' approach. You then have a great starting point to consolidate and improve teaching and learning in science 'Beyond the Classroom Boundaries'.

What are the options?

There are a large range of options open to us and this book could be viewed as the tip of the iceberg in terms of where the developments in outdoor learning could lead. The options open to you will depend on where you and the children are on the Escalator, the children's age and ability, school resources, grounds and support, and so on. An attempt has been made in Table 6.1 below to give some pointers as to what could assist the progression from one stage to another within the Escalator model. You will find references to

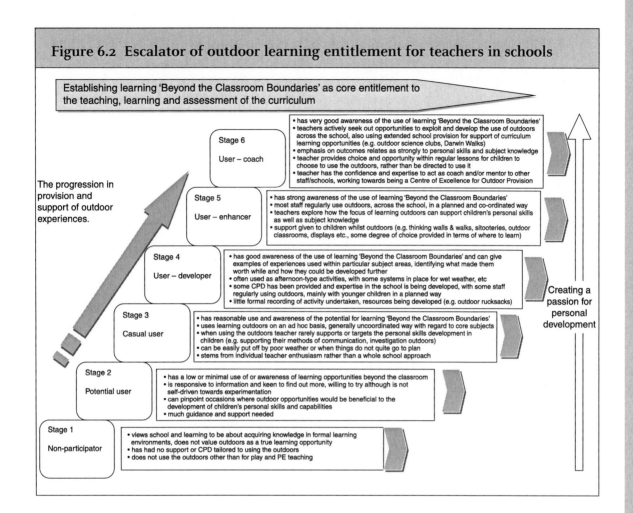

Figure 6.2 Escalator of outdoor learning entitlement for teachers in schools

Establishing learning 'Beyond the Classroom Boundaries' as core entitlement to the teaching, learning and assessment of the curriculum

The progression in provision and support of outdoor experiences.

Stage 6

User – coach
- has very good awareness of the use of learning 'Beyond the Classroom Boundaries'
- teachers actively seek out opportunities to exploit and develop the use of outdoors across the school, also using extended school provision for support of curriculum learning opportunities (e.g. outdoor science clubs, Darwin Walks)
- emphasis on outcomes relates as strongly to personal skills and subject knowledge
- teacher provides choice and opportunity within regular lessons for children to choose to use the outdoors, rather than be directed to use it
- teacher has the confidence and expertise to act as coach and/or mentor to other staff/schools, working towards being a Centre of Excellence for Outdoor Provision

Stage 5

User – enhancer
- has strong awareness of the use of learning 'Beyond the Classroom Boundaries'
- most staff regularly use outdoors, across the school, in a planned and co-ordinated way
- teachers explore how the focus of learning outdoors can support children's personal skills as well as subject knowledge
- support given to children whilst outdoors (e.g. thinking walls & walks, sitooteries, outdoor classrooms, displays etc., some degree of choice provided in terms of where to learn)

Stage 4

User – developer
- has good awareness of the use of learning 'Beyond the Classroom Boundaries' and can give examples of experiences used within particular subject areas, identifying what made them worth while and how they could be developed further
- often used as afternoon-type activities, with some systems in place for wet weather, etc
- some CPD has been provided and expertise in the school is being developed, with some staff regularly using outdoors, mainly with younger children in a planned way
- little formal recording of activity undertaken, resources being developed (e.g. outdoor rucksacks)

Stage 3

Casual user
- has reasonable use and awareness of the potential for learning 'Beyond the Classroom Boundaries'
- uses learning outdoors on an ad hoc basis, generally uncoordinated way with regard to core subjects
- when using the outdoors teacher rarely supports or targets the personal skills development in children (e.g. supporting their methods of communication, investigation outdoors)
- can be easily put off by poor weather or when things do not quite go to plan
- stems from individual teacher enthusiasm rather than a whole school approach

Stage 2

Potential user
- has a low or minimal use of or awareness of learning opportunities beyond the classroom
- is responsive to information and keen to find out more, willing to try although is not self-driven towards experimentation
- can pinpoint occasions where outdoor opportunities would be beneficial to the development of children's personal skills and capabilities
- much guidance and support needed

Stage 1

Non-participator
- views school and learning to be about acquiring knowledge in formal learning environments, does not value outdoors as a true learning opportunity
- has had no support or CPD tailored to using the outdoors
- does not use the outdoors other than for play and PE teaching

Creating a passion for personal development

other pages/sections in this book where activities, resources or experiences are explained in more detail.

Where to start?

Before moving blindly into the usual routines when starting an initiative, such as asking yourself, 'What can we do?', 'When shall we do it?', 'How much will it cost?', 'Who can help?', take a little time to think about what you are actually trying to achieve by expanding teaching and learning into work 'Beyond the Classroom Boundaries'.

Take the GROW approach that you have come across in each of the chapters in this book. Consider:

- What is my/our goal?

- What do you want to gain from this for the children's personal development? What do you want to gain for the teaching of science? What do you want to gain for yourself as a professional?

Kay's goals

Kay at Wheatlands Primary wanted to make working outdoors across early years (and indeed the whole of her primary school) the norm, where children would go outdoors to do their science in most weather. She also wanted children to work with increasing independence, which began in the Foundation Stage, and she was working with staff to ensure that the good practice was continued and developed as children moved into Key Stage 1 and then into Key Stage 2. Having this goal was important for Kay as it meant that she was developing a vision for the future which was shared with staff and children so that they could all work together in realizing the potential of science 'Beyond the Classroom Boundaries'.

Once you have identified your goals, and used the Escalators to consider your 'reality' you will be in a much stronger and more focused position to consider the options.

Now that you are at the stage to consider making changes, the key thing is to be open minded and not to limit yourself or your children. Create a table similar to that in Table 6.1 for yourself, leave your options open, then take some time to talk them over with colleagues. Do not 'just settle' for what may be easiest, but set aspirations that challenge, push practice forward and even take a few risks with your teaching and the children's learning.

Here are some suggestions:

Table 6.1 Moving forward

From Escalator stage	What teachers could do and pupils could be involved in …	Ideas/resources that could support
1–2	Identify the opportunities that are currently provided in the school grounds and the facilities/resources already available. Make a list of areas and locations that have potential of being improved or better tailored to outdoor learning. Consider the practicalities of working outdoors – what health and safety issues arise, what will the children need to be equipped with to work outside effectively etc?	See Chapter 2 See Chapter 4 to review the issues around managing the outdoor spaces and Chapter 5 to help address some of the health and safety safeguarding issues.

	Talk with the children about where they like to learn and why. Encourage the use of outdoor science equipment by putting out resources at informal learning times e.g. playtime, before/after school clubs. Give responsibility to children for looking after the equipment, showing other children how to use it, organizing it, etc.	See p. 83 to identify ideas such as Outdoor Science Boxes.
3–4	Review your current science planning for your year group or across the school. Does the current topic/theme plan link in with what is happening outdoors e.g. life cycles in the spring term, habitats in the summer term. Have you set up your outdoors Science Role-play Shed and built it into topic planning?	See chapter endings to see examples of how to use the school grounds in science See Chapter 9 on Science Sheds
	Explore what CPD opportunities are available to yourself and your staff. Consider who may be best at leading this area of development with you – maybe a colleague from another Key Stage, a teaching assistant, group of parents, a local scientist, etc.	
	Consider what visual and verbal protocols would support children learning outdoors.	
	Observe children working in pairs, small groups and teams during structured activities outdoors. Consider which of the areas of Personal Capabilities they are strong in and which you feel they need support with. Talk with them using simple AfL strategies, such as thumbs-up-sideways-down about how they feel when working in a different type of space and what type of things would help them learn better	See Chapter 3 to read more about the Personal Capabilities children use when working in this way, and how simple strategies can support their behaviour and learning.

(Continued Overleaf)

Table 6.1 Continued

From escalator stage	What teachers could do and pupils could be involved in . . .	Ideas/resources that could support
5–6	Discuss with colleagues how the range of outdoor learning opportunities can be refined to enhance the development of children's Personal Capabilities. Consider when it would be appropriate for children to be given greater autonomy and choice in deciding for themselves when to work within the classroom and beyond it.	See Chapters 3, 4 and 6. Specifically consider if any of the teacher ideas, could be of inspiration to you
	If aspiring to move forward the whole school, consider how all staff can work towards using the outdoors to support science learning. What could the whole school be involved in (e.g. setting up a Darwin Thinking Path, Science Outdoor Story Sacks)?	See Chapter 8 – Darwin's Thinking Walk

See Chapter 10 – Science Story Sacks |
| | Consider if there are any approved adults who could work with children (e.g. visiting artists making Darwin Stepping Stones, or working as 'Expert' in the Science Outdoor Role-play Sheds) | See Chapter 8 – Darwin's Thinking Walk

See Chapter 9 |

What will you do?

So maybe we have arrived at the million-dollar question: What are you actually going to do?'

You are clear on your **Goal**, you have reviewed where you think your school, teachers and children are, your **Reality,** and you have read about and decided on the range of **Options** open to you. So what next?

Now you are at the stage of deciding the '*What*' as in 'What will we do?' Do you make the final decisions on your own, with a colleague or even with the pupils? However you decide to do it, the key thing is to be ambitious and to challenge yourself while having a clear sense of how it can be achieved, in what timescale, who would be involved, and how you will measure success. After all, you will want to share your success and be able to identify what worked well and the real gems of experience that help to tell the story. Indeed, to understand why things have and did not work is essential for you as a professional learner and curriculum developer.

You may consider some of the following as ways to catalogue what has been happening and its effects:

- *A big book*: to record progress in pictures accompanied by children's comments.
- *A working wall*: an interactive, collaborative display where all groups in the school can display photographs, pieces of work, lesson ideas, reflections; just think if this could even be outdoors!
- *Outdoor working party*: group of children (similar to the School Council) who advise on outdoor learning activities, review experiences, take the lead in organizing outdoor equipment, provide a forum for reflection and undertake pupil surveys. Children could produce a 'Newsletter' communicating progress to other children, teachers, school governors and parents. In primary schools the older children could 'mentor' or 'buddy' older Key Stage 1 children in these types of activity, and certainly could 'interview younger children about their science outdoors, how they used the Science Haversacks and Outdoors Science Boxes or even how they created 'Jack's Bean Garden' (see Chapter 9, Science Outdoor Story Sheds)
- *Assemblies and reward certificates*: focusing on 'What Progress We've Made' learning outdoors.
- *The school website*: with photographs, news items and information for parents of forthcoming activities and events 'Beyond the Classroom Boundaries'.

Tables 6.2–6.6 are examples of frameworks that you could use to guide your planning, within one class or across the school.

The real power behind any of the suggested frameworks, or indeed the use of the Escalator is, of course, the level of honesty and objectivity of the user. In many cases it will be necessary to broaden one's understanding of what is actually happening in other classrooms in the school. Take this opportunity at this point in your reading of this book to stop and begin to objectively judge the stage of your own work and that of other teachers, and to be frank about the true impact this is having on pupils. Talk to teachers, talk to pupils, it is through such talking processes that you as an innovator will really glean a sense of the status quo in your school.

Table 6.2 A whole-school Half-termly planner indicating science topics and whole-school developments 'Beyond the Classroom Boundaries'

	Foundation/ Reception	Y1	Y2	Y3	Y4	Y5	Y6	Whole school
Autumn 1								
Autumn 2								
Spring 1								
Spring 2								
Summer 1								
Summer 2								

Table 6.3 A year group monthly planner indicating general provision across the curriculum 'Beyond the Classroom Boundaries'

Month	Learning experiences	General resources (including people)
September	Science – Literacy – Numeracy – Etc.	
October		
November		
December		
January		
. . .		

Table 6.4 A planner for teachers to use indicating short-term and longer term goals

	Quick Win 1	Quick Win 2	Longer-term goal
What is your goal?			
What will you do?			
What will you need?			
How will you evidence this activity?			
How much effort will this require? (1–5 scale)			

Table 6.5 Using the GROW approach to map the experience

Goal What is your Goal?	
Reality What is your Reality?	
Options What are your Options?	
Will What Will you do?	

Table 6.6 A written reflection grid for teachers

1 Describe the lesson/activity that took place 'Beyond the Classroom Boundaries'

2 How did you feel about the process and outcome?

3 What was good and what needed adjusting about the lesson/activity?

4 What have you learnt? What do you take away from this lesson/activity?

5 If you did it again what would you do?

Practical ideas for teaching plants 'Beyond the Classroom Boundaries'

'Beyond the Classroom Boundaries' offers a range of opportunities to develop children's understanding of plants as well as the prospect of helping to progress children's Personal Capabilities. In this section of the chapter we have sought to provide activity suggestions for children engaged at each stage of the Escalator to help them and indeed to support staff in moving children forward.

Stage 1: Non-participator

Children have no exposure to experiences learning beyond the classroom; they do not see the value, purpose in learning outside; they do not take initiative to go outside, showing unwillingness at times; they adopt the view that learning is about acquiring knowledge in formal settings, and that the outdoors is just for play.

For these children early steps to encourage them to be outdoors are key. Try short, structured tasks: for example, in pairs, they go and photograph as many plants as they can find in 10 minutes. They can then be asked to select their two favourite pictures and to create a display by encouraging them to mount the photographs outdoors, perhaps on a fence, using plastic wallets and placing keywords related to the plants inside or around the wallet using waterproof markers.

This is also a good point to introduce the 'Outdoors Science Boxes' to children. Take the class outside and show them where the boxes are and take out all the equipment and encourage children, in groups, to think about what the equipment can be used for and what they might do with it during play and lunchtimes. Also ask children to think about the rules that were explained to the whole school in assembly, when the boxes were first introduced to the school.

Over the first couple of weeks do have 'share times' for a couple of minutes after playtimes so that children can talk about what they have used and what they did to celebrate and value children using the boxes.

In pairs, supported by a teacher, assistant or parent helper children could explore the school grounds with the aid of an identification key. They could take part in a plant hunt where they have to spot as many different varieties of plants as they can. If the identification keys are laminated then the children could use a whiteboard marker to put a tick or a smiley face beside the ones they find.

Stage 2: Potential user

Children have a low or minimal use or awareness of learning beyond the classroom; they show enthusiasm but are unaware of how to harness this to enhance learning; they may 'hover' and watch others, unsure of the protocols outdoors; they are responsive to information and keen to find out more, willing to try; they pinpoint occasions where outdoor opportunities would be beneficial to them; they talk about the constraints of the classroom and the freedom outdoors.

When beginning science activities discuss with the children whether they think this activity is one that would be better to do outside rather than inside. For example, in plant topics we often grow seeds and answer questions such as:

- Which plant grows the tallest?
- Where is the best place for the plant to grow?
- How much water should we give the plant?
- Will the plant grow better if we feed it?

We could also ask which kind of container could we put our seeds in, and offer children a range of containers from plant pots, to old metal teapots, old boots and so on. Then ask the children to think about how they would organize themselves to carry out this activity outside, for example:

- What will they need? (Equipment could be placed in their Science Haversacks)
- Where will they plant their seeds/plants?
- How will they label their seeds so that they know which seeds/plants are part of which test?
- Will it be all right just to use one seed? What happens if it dies?
- How could they show how the plant grew from seed to when it flowered?

This is a great way to begin, by presenting children with a more structured activity, which is mostly outdoors. It also means that as long as the area is accessible at play-

and dinner times, children might be interested enough to visit the plants and see how they are progressing.

Stage 3: Casual user

Children have some awareness of learning outdoors, requiring a lot of structure to work effectively; adult intervention is required on how and where to best use resources that are supplied; have reasonable use and awareness of the potential for learning 'Beyond the Classroom Boundaries'; some children are involved in guiding others in informal settings, for example, playtime activities; have access to and use outdoor resources, organizing and managing the resources in groups, pairs or teams – for example, science boxes, during non-formal learning times during playtime or before/ after school activities.

Here the children should be comfortable with using the Outdoors Science Boxes for their own explorations and enjoyment. The aim here is to introduce children to a range of activities that they can then repeat on their own during play- and lunchtimes. In this activity we suggest carrying out a 'Smart Hunt' modelled on the materials from *Smart Science* (see www.personalcapabilities.co.uk) where children:

- work in teams of three
- delegate roles within the team (e.g. recorder, timekeeper, photographer)
- think about health and safety issues (e.g. look after each other, do not eat plants).

The teams are directed to different areas in the school grounds to explore with their senses the variety of plants that they can find. This activity encourages increasingly independent work, in a semi-structured way, assisted by group roles and suggested response sheets and timescales. Building autonomy at the 'Casual User' stage will prompt ownership and personal responsibility for learning outdoors. The children in this activity will be asked to:

- work in the given area
- use equipment such as cameras, hand lenses, rulers and tape measures, identification sheets or books
- locate and identify the plants in the area that they have been given
- log the different plants in that area using a digital or video camera
- record where the plant is growing, any animals living on or around the plant, and what it feels and smells like. They could use an Easispeak microphone.

Back in the classroom the groups can then work in pairs to research information about the different plants which can then become, for example, part of a big book, or a set of fact files that might be placed outdoors.

Let the children know that if they want to use the hand lenses and identification books or charts during play and dinnertimes they can and if they find any new plants to add them to the fact file or big book.

Stage 4: Developer

Children are aware of expectations for behaviour and learning requires structure; suggest when outdoors may be best; they require some guidance and support, less so when working in collaborative groups; are developing a range of methods to note and record learning outdoors; they give examples of experiences, identifying what made them worthwhile; they take up opportunities to work outside when offered; they are involved in outdoor learning clubs (e.g. gardening club).

Encourage pairs to make a list of questions that interest them about plants, for example:

- Which are the tallest plants in our school?
- Where are the most plants found?
- How many different shades of green can be found among the plants in our school grounds?
- Which plant takes up the most space?
- Which plant is the oldest? How can we find out?

By encouraging such questions using question stems, such as how, which, why, who, when, does, what, children can then be scaffolded into asking different questions.

If the school has a vegetable garden area then children could become 'Young Horticulturalists'. Give them a badge with this title on or get them to make their own; they will really love wearing it and explaining to others what it means.

Ask children to research what a horticulturalist is and does. They should find out that horticulture is all about plant science, how to grow plants and cultivate them. Encourage the children to become more proactive by asking them what they would like to plant; some children might need the adult to offer plants to choose from, which could include:

- vegetables
- flowers
- plants that smell
- plant that are interesting (and safe) to touch
- plants that make interesting sounds in the breeze.

Stage 5: User enhancer

Children have strong awareness of the use of learning 'Beyond the Classroom Boundaries'; they plan projects to create something or investigate something using outdoors at appropriate staging points; they are able to organize themselves when working outdoors, asking for assistance when necessary; they take ownership of the outdoors by making suggestions for learning outdoors, the types of resources and possible changes that could be made to improve experiences; they use resources wisely when outdoors, aware of the need to store them well.

Some children might be encouraged to take on the mantle of expert as Young Ecoconsultants. They could consider what changes or improvements could be made to the school grounds and manage a small project, for example:

- Finding out how many different animals live in or visit the school grounds. The children could keep a chart on a white- or chalk-board in the school grounds, so that everyone can add to it, when they have seen a new animal.

- Children could plan and help to create a senses garden or plant a trough in their own area or somewhere in the school. This might tie in with early years topics on the senses and once the plants have been planted the children could organize a 'garden opening ceremony' where they invite someone to 'cut the ribbon' to open their garden, or perhaps give a presentation about their new garden (however big or small!) during a school assembly to the rest of the school.

Stage 6: Coach

Children have very good awareness of the use of learning 'Beyond the Classroom Boundaries'; they recognize opportunities to extend their learning independently; they talk about and describe the benefits and protocols of learning outdoors and indoors; they are proactive in making suggestions about what they would like to do and have outdoors; they exploit the outdoors in a mature manner, needing limited guidance on behaviours for learning outdoors; they actively support peers when working outdoors, be they the same age and younger.

There will be children in the school who could be designated as 'School Grounds Ambassadors' and be asked to take visitors around the outdoors area to explain how they used it in science, different areas, plants and so on. There is no reason why early years children cannot do this, either in pairs or with a child from an older class or an adult working with the children as a 'chaperone'. In terms of Personal Capabilities this is an excellent approach and many of the project schools are working towards developing pupils confidence so that they are able to be placed on a rota of 'School Grounds Ambassadors.' If you have ever been a visitor to another school or classroom you will know how eager children are to share their experiences and learning with you. This could of course be at a point when new children are visiting the early years prior to the new academic year, older children could be asked to 'buddy up' with a new starter to show them around the school grounds and tell them about what lives there, use the Outdoors Science Box with them. Again there is no reason why a whole class cannot be involved in this with, for example, Key Stage 1 children showing new starters into Foundation around, under the guidance of teachers from both classes. It would also mean that when the new starters began school they would see friendly faces and could buddy up for other science activities across the year including the older child showing younger children how to do things, such as plant seeds, in the Science Role-play Shed (Chapter 9) or in reading Foundation Stage children stories and sharing activities from the Science Outdoor Story Sacks (Chapter 10).

7

Resourcing science 'Beyond the Classroom Boundaries'

What is the goal?

Resourcing science 'Beyond the Classroom Boundaries' is quite easy; the important part is what we hope to achieve by collecting or purchasing a range of equipment to support science outdoors. If we return to the basic philosophy behind science outdoors, this will help us to develop appropriate resources and associated practices. We are aiming to:

- use the whole school outdoor environment to access the primary science curriculum
- develop children's Personal Capabilities.

A carefully selected range of resources for outdoors can help to motivate children to explore activities either on their own or with other learners. In choosing resources the teacher should aim to provide items that do not require an adult to act as mediator. In this way children are encouraged to work independently outdoors, making their own decisions about whom, how and what they work with. As children become more experienced and confident in using the resources they will become self-regulating and increasingly motivated to return again and again to explore the outdoors when they see it is appropriate and of benefit to their learning. Of course, this is a highly ambitious target, but one to which we should all aspire.

Do not forget that children enjoy sharing their experiences, ideas and products, so strategies to support communicating and developing a positive self-image in working outdoors are vitally important in enriching the development of science knowledge and understanding. For example, it is useful to ask questions such as:

- Which resources will be supportive of learning and provide novelty and added stimulus to learning?
- Which resources will challenge the children to show their learning in creative ways?

After all, being outdoors should be fun, something special, as well as educationally challenging.

What is the reality?

The reality is that we can have all the resources in the world but it will not necessarily lead to good science. When Megan (aged 6) was asked about using the outdoor science boxes in her school she answered,

> We have not used it because we are not sure whose turn it is, which class can use it.

So, what did Megan need in order to be able to confidently use science resources outdoors independently? What Megan really needed was to:

- have access to a wide range of resources for science
- know which resources she has access to
- understand how to use scientific equipment safely
- know that she is allowed to choose and use her own equipment when engaged in science
- understand that if she gets equipment out she is responsible for looking after and returning it.

The reality is that unless Megan is taught how to use different kinds of science equipment and the procedures that go with them, she is less likely to use them to their fullest potential in creative ways.

When children are offered different kinds of equipment and given guidance in choosing it and using it they are more likely to be confident and increasingly independent as opposed to when this has been predetermined for them.

Indeed, some interesting choices can also be made, such as those made by Aiden (aged 5) who likes to look for bugs. He chose to use magnifying lenses, Bug Huts and also something to offer him comfort while exploring.

> I use the knee pads when I look for bugs because it stops our knees from getting achey.

Access to science resources should be like accessing an 'Aladdin's Cave' for children: lots of wonderful, interesting things to use, where new things keep appearing across the school year. Children should be able to use science equipment not just in more formal science activities but also during play- and lunchtimes so that their interest in science and the world around them is nurtured outside of the formal curriculum.

It is useful to now consider the extent to which the above scenarios match your own school practices. How close are you to children having this kind of access to resources outdoors?

What are the options?

First of all common sense says we do not necessarily need lots of *new* equipment: sometimes it is simply the reorganization of existing resources and a few new purchases to enhance science outdoor provision. Starting off by auditing resources is really useful and project teachers who did this realized how much was already in school that can be used outside. The key questions that they asked were:

- What resources do we already have and do we need to duplicate any?
- How can we best use the resources that we already have to support science?
- What other kind of resources could we collect or purchase that would challenge and enhance and offer challenging experiences in science?
- How can resources be organized to support and develop independence?
- What kind of resources would encourage creative exploration?

An audit of resources sets out your starting points for thinking about how you would like to use and organize them to extend science learning 'Beyond the Classroom Boundaries'. Early years (Foundation Stage, 3–5 year olds) are used to organizing resources for children to use outdoors, but sometimes these do not have an obvious science outcome, or access to scientific equipment is limited or, at the best, linked to specific topics and therefore out only during that topic. For later early years classes (5–7 year olds) creating outdoors areas and obtaining similar equipment to Foundation Stage will be important, so sand and water trays, water cascades, bicycles and cars as well as outdoor construction materials should be the norm.

In this section our aim is not to duplicate what most schools know about early years outdoor equipment but to suggest some quite different approaches to introducing children to science specific resources.

Outdoors Science Box

A useful first step in resourcing science 'Beyond the Classroom Boundaries' is to develop an 'Outdoors Science Box', this proved a great hit with project staff and children. The Outdoors Science Box is a large plastic box containing all sorts of resources, which is placed outside during the school day for children to use, mainly at play- and lunchtimes.

This idea came from Sue Harrison at Cheveley Park Primary School in Durham and was designed to provide support for children to manage their own learning in both lessons and during playtimes. Such a resource can enhance the opportunities for children to develop self-management skills and to:

- explore their scientific understanding of their immediate environment
- make their own choices about what they want to do
- use a range of scientific equipment safely
- record and communicate their experiences in a range of ways should they wish to do so
- organize, plan and take responsibility for how they will carry out a task
- think about how they learn as well as what they learn.

Sue's Outdoors Science Boxes

Sue's idea was to have one or more large plastic trolley boxes that could be easily pushed by pupils. Sue developed the idea of outdoor 'Science Monitors' who took charge of these boxes and whose responsibility it was to bring them back in at the end of the day.

In using this approach Sue considered health and safety issues and made sure that children were not lifting or carrying heavy boxes and that the contents were safe and kept dry by storing them in a weather-proof box.

Using a monitor system ensured that children were responsible for resources outdoors, by giving them the role of checking the contents and retrieving any items left out; this also meant that staff did not have the added burden of managing the resources on a daily basis.

The children were trained in looking after the resources and informing Sue when items had been lost, damaged or needed replenishing. A rota ensured that different children across the school were given the opportunity to be in charge of these resources.

Personal Capability self-management: to be responsible for the things around me.

Sue's idea provides opportunities for children to feel that they are in charge of what they do by taking responsibility for what they need, how it is maintained and managed. This necessary life skill is what many parents often strive for at home and in-school experience can only complement such endeavours.

This idea was taken on board successfully by a number of the project schools: for example, Carol Sampey at Shaw Primary School developed boxes for early years and Key Stage 2. The early years children were taught how to use the equipment and given ideas on how to use the resources during their play and lunchtimes. It did not take long before the boxes became an integral part of playtime life as well as being used during more formal teacher-initiated activities. Katy (age 6) shows her appreciation of the boxes:

> There is a science box for Key Stage 1 and one for Key Stage 2. There's lots of science stuff, me and Alice were playing with them looking for bugs and insects. We used a magnifying pot with a lid. It was really fun. Playtime is better when we use magnifiers and binoculars, we saw lots of things far away, we see lots of birds.

What is interesting here is that Katy understands that there are different boxes for both Key Stages and that she can use them whenever she wants, which she obviously does, and in doing so is also showing that she can choose and use appropriate science equipment for the job she wants it to do.

In chatting informally with Aiden (age 5) we are left in no doubt that he knows what binoculars are for and how to use them.

> Jacob keeps getting the binoculars out, to look for the girls who are trying to catch Joe.

For some children using the outdoor boxes introduces them to new experiences and allows them to repeat activities that they enjoy; something that is not always possible during timetable lessons. It is amazing how curious children are and how determined to find what they are looking for, and how what they learn informally is then transferred to different contexts, in Jennifer's (age 6) case, her home.

> When I had the magnifying glass and a little box to catch beetles, with my friend Emily, we saw a ladybird and we saw woodlice. That was the first time I saw a woodlice, when I am at home I keep seeing them now.

When putting science boxes together include paper, pencils, crayons (not felt tips), plastic wallets, pegs and string, so that when they want to children can record their experiences and then put their work in plastic wallets to display on a washing line, or place on a fence or railing. Do suggest to children that they use some Sellotape to tape up the end, then, hang the wallet upside down, so that their work remains waterproof.

I like the science boxes because it is fun because you can draw, and there are plastic wallets you can put a picture in them and peg them on the washing line. Maddie (age 6)

Outdoors Science Haversacks

Teachers interpreted the Outdoors Science Box in different ways. In this example, Sarah from Castleside Primary decided not to have an Outdoors Science Box but Outdoors Science Haversacks. She thought that these would:

- motivate children to want to work outside
- encourage children to take responsibility of resources
- allow individuals and small groups of children to work independently
- organize, plan and take responsibility for how they will carry out their tasks
- think about how they learn as well as what they learn.

The 'Outdoors Science Haversack' concept was also meant to offer a flexible approach for staff, who, if they wanted to could change the contents of the haversack to different science activities outside.

An 'Outdoors Science Haversack' is carried by the children whenever they work outside. One person in each group takes the haversack out with them and is in charge of checking the contents prior to going out, and then before returning to the classroom to ensure that everything on the laminated checklist is present.

Sarah changed the haversack to suit the class science topic. She included the following for a haversack on invertebrates:

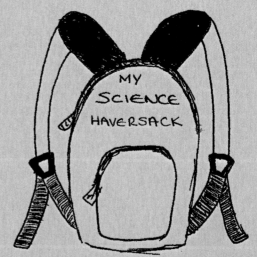

- digital camera
- Easispeak microphone
- magnifiers
- clipboard
- paper/notebook
- pencil
- sharpener
- pooters/collecting pots
- identification keys for invertebrates and plants.

7.1 Science Haversack

Like the Outdoors Science Box the contents of the haversack can be changed to include a focus on a particular aspect of science learning; for example, changing the haversack to an 'Outdoors ICT Haversack'. This could contain equipment for collecting and recording data, ways of working and children's ideas such as:

- a portable digital microscope
- Easispeak microphones
- digital camera
- video recorder
- Talk Buttons.

Sarah's Science Haversacks also served a different purpose to the Outdoor Science Boxes. Unlike the boxes, the haversacks were not left outside all day, although the children could take them outside at play- and lunchtimes if they requested. The haversacks were developed to help support children working independently on specific areas of science during lesson times. The children would either pick up a 'ready prepared' haversack, which were best when introducing children to the idea, or they would use the 'basic' or 'standard' haversack, which contained pencils, clipboard or notebook, magnifying lens, Easispeak microphone and/or camera. If using the 'basic' version, the children also collected specific equipment that they thought they might need for the activity. By using both approaches, the children were scaffolded into developing independence in choosing, using and being responsible for the equipment that they use.

For ideas of what kind of science activities could be put into the haversack, see the final section of this chapter.

Bethan's Science Trolley

A different approach was suggested by Bethan's (Head teacher at Framwellgate Primary). Her response was to suggest using trolleys from the Foundation Stage so that the children could put their equipment on the trolley and take it to where they wanted to work. The trolley is really useful when children want to move larger items of equipment around for their work in science.

Outdoor chalk- and whiteboards

The range of ideas and practice that emerged from project schools was exciting and very practical. For example, some schools decided that children should have access to chalk- or whiteboards on outside walls so that children could use them to record ideas, questions, observations and even draw designs or sketch birds they had observed.

Wall-mounted chalk- and whiteboards

Grenoside Primary School planned to purchase wall-mountable whiteboards for outside in the school grounds, which when placed at a suitable height would allow children to use them to record ideas, results of investigations, sketch, record observations of birds or invertebrates or even write down their own questions.

Of course, in the early years recording often challenges both children and teachers; however the whiteboard encourages mark-making, and allows children to feel they can draw pictures or attempt words and sentences without the worry of having to handle clipboards, pencils, and so on. Such a technique also offers a more collaborative approach to recording.

Sound Mapping

Teachers at Bradway Primary School described how children in the early years used coloured whiteboard pens when Sound Mapping their school. Children stopped and listened to the sounds of the world around them and made marks that represented the sounds, and even chose colours that they felt indicated the type of sound, not on wallmounted boards but individual hand-held whiteboards.

Teachers suggested all sorts of creative approaches. In the next example, Sue at Cheveley Primary was keen to support not just the children but also teachers and other adults. For this reason she displayed activity ideas around the school grounds, which served the dual purpose of engaging children in activity and reminding teachers (and other adults) of the potential for science of different areas in the school grounds.

Displaying activities outside

Sue used visual cues by taking photographs of children engaged in activities such as colour matching in nature, using pooters to collect invertebrates and binoculars to spot birds. It was also important to include activities from other areas of science such as sound, materials and forces: for example, 'What can you hear?' stations, 'How many different materials can you find in this area?' and 'What kind of forces are in action when you play on the climbing frame?'

She supported children by making sure that each activity had text and pictures, which meant that a wide range of age

7.2 How many shades of green can you find?

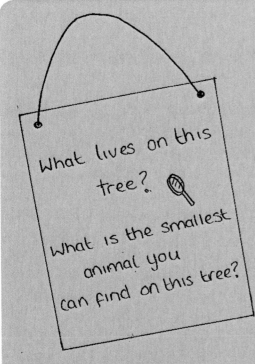

7.3 *What lives on this tree?*

7.4 *What kind of sound can you make?*

and ability levels could access the activities independently, without the need for adult help or supervision. At the same time the suggestions provided teachers with starter ideas and activities for their science topics. This also meant that if children wanted to they could have a go at the activities she had placed around the school grounds; they could do this during play and lunchtimes as well as being directed to specific activities during timetabled science lessons.

7.5 *Use binoculars to observe birds in the school grounds*

Being given options and having choice of what, where and when science learning takes place enhances children's self-directed learning. For some the simple act of watching others will be enough to encourage them to take independent strides for themselves. Encouraging autonomy in these simple ways, by freeing up access to resources, provides useful stimulus for children to develop a sense of autonomy in their learning. They begin by trying out what is provided and rapidly move off to experiment on their own.

What will you do?

The most successful resources are those that are part of a well-thought-out action plan based on the resources already available and the key goals for developing science outside the classroom boundaries.

- What are you trying to achieve by developing resources boxes or haversacks to support working 'Beyond the Classroom Boundaries' in relation to science and also Personal Capabilities, such as self-management and creativity?
- What system of managing resources for science outdoors would best suit your children and staff?
- What do you already have available in school that could be used in the resource kits?
- Which resources would you, your staff and the children like you to purchase?

Before purchasing any resources think about:

- asking local businesses for support e.g. clipboards
- using friends of school to fundraise
- sending letters home to see what parents can donate e.g. haversacks.

Most importantly once the children have begun using the resources and understand what they are for and how to use them, ask them what they would like in the boxes and haversacks. You might be surprised, as we were, when Maddie (age 6) and Sam (age 6) made these suggestions:

> I would like to have an art and craft box we could make things to put on the washing line and make bugs. (Maddie)

> I would like some fossils in the box, pretend ones or real ones. (Sam)

The suggestions we have offered here of course should complement children being able to engage in exploration on their own terms. We should not try to control how children use the resources; the intention is that approaches such as the Outdoors Science Boxes, Science Haversack and Trolley are to encourage and support children's experiences. Children do need opportunities to follow their own interests just like the children in Foundation Stage did at Bradway Primary School.

Mud, glorious mud!

The best science is often child-initiated, as happened one day when there were lots of puddles around the playground. A child picked up a small spade and began to see if he could connect one puddle to another to let the water flow between them. Teachers spotted this and encouraged him. Soon other children became involved and joined in and teachers used this great context to develop language, talking about dry, wet, liquids, changes in textures and colours. Of course, the moment came when a child added soil, and exploration of mud followed. Messy but wonderful!

Practical ideas for Science Haversack for children to use 'Beyond the Classroom Boundaries'

In this section you will find ideas for the contents of a range of themed Science Haversacks. These are only suggestions and hopefully they will spark off lots of other even better ideas. Included are suggestions for activities, but what would be great is if you create the Science Haversacks that include an activity card which has a photograph suggesting what the activity is alongside word explanations, so that it can be used by different abilities.

All of the equipment suggested was available on the TTS website at time of print (www.tts-group.co.uk).

Table 7.1 Ideas for Science Haversacks

Science Haversack name	Science Haversack contents	Activity suggestions	Personal capabilities
Fill your matchbox against the Clock	• Match box or larger container • Sand timer or electronic egg timer	Children are given items to collect from outdoors against the clock. The collecting container will depend on the size of the objects the children have to collect. For example: • Ten things that are different shades of green or brown, or yellow, etc. • One of each of these, something rough, smooth, flexible, transparent, hard, soft, opaque.	Self-management – *to be organized* The fact that this activity is against the clock requires the children to focus their attentions and organize themselves. Working in pairs or threes encouraging the children to delegate, communicate and plan ahead for where they will search.

(Continued Overleaf)

Table 7.1 Continued

Science Haversack name	Science Haversack contents	Activity suggestions	Personal capabilities
Make Your Own Senses Trail	• Plastic arrows • Senses signs: LOOK, SMELL, TOUCH, LISTEN (words and symbols)	Children create an interesting trail for the rest of the class to follow, using the items in the Haversack. When completed the children go on the Senses Trail and report back to the teacher and group that set the trail. The trail can be changed so other groups have different tasks e.g. Minibeast Trail, Magnets Trail, Textures Trail.	Creativity – *to be imaginative* Whatever the contexts children can be given some free reign and autonomy to share ideas and see them through. Encourage children to be thoughtful and imaginative, to set challenges for their peers and to be a little out of the ordinary.
Water and Holes	• Water trough • Watering can • Anything with holes in e.g. tea strainer, colanders, sieves, plastic bottles with holes punched • Camera	Children use the watering can to pour water through the different containers to see what happens e.g. pattern of the water, time it takes to flow through. If children work in pairs, encourage them to take photographs to share with everyone on the interactive whiteboard.	Communication – *to describe and explain* The use of the cameras will provide visual stimuli for the children to structure the way they talk to their peers following this task. Discuss with the children that people often use props (e.g. pictures, objects, photos, posters, etc.) to help them tell stories.

What can you see?	Science Haversack containing: • 2-way magnifier • Magnifying lenses • Magnifying pots • Frame magnifier, • Fresnel Lens • Pocket microscope	The haversack could contain one or more of these types of magnifier. If children are familiar with one type you could introduce them to something more powerful or sophisticated. Children could have free choice of what to view or given specific items, such as soil, grass, leaves, flower heads, invertebrates and tree bark.	Perseverance – children take time to locate objects and use different magnifiers, not giving up if the image is not clear first time.
Toys	Jumping frogs and other moving toys	Children explore how the toys move, which surfaces they can move over, how far the can move.	Teamwork – *to work well together* Discuss with the children what it will take to work well in a team. Encourage clear roles e.g. tester, recorder, presenter, organizer, collector, maker.
Making a dark, dark den	• Blackout blankets/ fabric • Torches	Children are challenged to make a den in the school grounds which is very dark, so dark that they will need to use the torches inside.	

(Continued Overleaf)

Table 7.1 Continued

Science Haversack name	Science Haversack contents	Activity suggestions	Personal capabilities
Scarecrow	• Pencils • Sketchpad • Photos of different scarecrows • Elastic bands • String • Fabric pieces • Straw	Children design and make a scarecrow for the vegetable patch or flower bed 　Children can choose to use natural objects that they find around the school grounds and resources from inside the haversack to make their scarecrow.	Creativity – encourage children to come up with new ideas and share them with peers.
Parachutes	Ready-made parachutes or material squares with holes in corners, string, washers or small plastic person	Children make and explore launching parachutes from different places in the school grounds.	Teamwork and organising themselves to solve problems.
Sensing sound	• Elephant Ears (See Chapter 1)	Children listen to the natural sounds around the school grounds 　They can either use an Easispeak microphone or clipboard and pencil to suggest what the sound feels like and what colour they think would represent the sound. If writing they could make a mark e.g. shape or squiggle to represent the sound.	Communication – *to listen and respond* 　Develop listening skills with this task. Use simple activities before the 'ears' to focus the children in on listening carefully e.g. play Chinese Whispers and emphasize the need to be alert to sound and to respond quickly to capture what they hear.

WOW sounds	• Space phone • Thunder tube	Children explore these great pieces of sound-making equipment (see TTS online catalogue www.tts-group.co.uk).	Creativity and problem solving – to be curious to choose questions to explore.
Bumpy Rides	Toy cars of different shapes and sizes	Where is the roughest, smoothest, steepest, bumpiest place for the cars to travel?	New equipment will undoubtedly stimulate curiosity. Talk to the children about what it means to be curious and encourage them to ask questions using question words – what, why, where, when, how, etc.
Land Yacht Race	Land yachts (balsa buggies with sails)	Children race the buggies using hand fans to move the buggies. *Yachts*: What could make the yacht faster? How can we make the yacht move more smoothly? Etc.	Problem solving e.g. Why doesn't the wheel go round? What should I do?
Looking underwater	• Marine viewer • Water trough • Variety of different objects or plastic underwater animals	Children explore what they can see underwater using the marine viewer in the water trough.	Working together – two taking with someone else.
Mini kites	Mini kites	Children explore forces and movement of the kites.	Perseverance if the kite does not work first time.

(Continued Overleaf)

Table 7.1 Continued

Science Haversack name	Science Haversack contents	Activity suggestions	Personal capabilities
Colour the world	Coloured acetate sheets or outlandish glasses with coloured acetate	Children explore what the school grounds are like looking through glasses with different coloured lenses.	Work with a friend and describe/share what they can see, comparing their responses and developing communication skills.
Seeing round corners and over walls	Periscope	Children explore looking around and over objects using the periscope.	Self management – working on own.
Dark secret places	Torches	Children use the torch beam to highlight what they can see in dark secret places such as under a stone, in a pile of logs, on a tree trunk.	Working together for example one child holding up stone, one holding torch other observing.

Seeing teddy	Teddy reflective strips	The children find out which reflective strip can be seen best when Teddy is out and about in the school grounds.
		    You could include a story about Teddy wanting to be seen when walking home from school.
			Problem solving – to work things out.
			    Solving problems can take different forms – from trial and error to planned actions. Before the children set off on this task, encourage them to have a few ideas about what they will do, how they will do it and how they might record what they discover. Give assistance where appropriate.

Magnetic Hunt	• Giant horseshoe magnet
	• Magnetic strips
	• Alnico magnet
	• Magnetic wand
	• Magnetic toy cars/train
	• Tape measure
		Find out what is magnetic around the school grounds. Take a photograph of magnetic objects.
		    How far can you pull the magnetic train engine and wagons before they break apart?
		    Which is the steepest hill you pull the magnetic train engine and wagons before they break apart?
		    Which is the bumpiest place you pull the magnetic train engine and wagons before they break apart?
			Self management and communication.
			    Children taking photos or video chips of what they are doing, to share back in the classroom with other children and the teacher.

Panning for gold	• Pieces of iron
	• Pyrites (fools gold)
	• Sieves with different sized meshes
	• Water tray
		Some rock collections have samples of iron pyrites, you might also be able to get some from local geologists and university departments.
		    It is called fools' gold because it looks like gold, but isn't. Children use the sieves in the water tray to sieve sand to find the iron pyrites.
			Self-management – to have stickability
			    This task could test the patience of some children, therefore ask them to work in pairs and to consider how they could overcome frustration if they don't find 'gold' quickly, or how they could share out the task so that they are successful in the end.

Darwin's Thinking Path 'Beyond the Classroom Boundaries'

What is the goal?

Many of the schools involved with this project began exploring science 'Beyond the Classroom Boundaries' were also celebrating the two-hundredth birthday of Charles Darwin. This sparked the imagination of both teachers and pupils and led to the development of a 'Darwin Thinking Path' that was based on Darwin's 'Thinking Path' which he created at Down House in Kent where he and his family lived, and where of course he wrote his most famous books On *the Origin of Species* and *The Descent of Man*.

Darwin created a path that followed the outskirts of the grounds of Down House, passing by hedgerows, a field and through woodland. Darwin's life was dominated by routine; he walked the path every day for an hour from midday, enjoying the physical exercise and solitude that allowed him space to think. His use of the path tells us a lot about how Darwin as a scientist worked; each day he would take time out to think, undisturbed, as he walked round his Thinking Path.

8.1 Down House

No matter what the weather might be like he went for a stroll at the Sand Path with his dog Polly, a white terrier. Along the way he would often stop by the greenhouse to check up on how his plant experiments were doing.

(www.aboutdarwin.com/darwin/CD_Daily.html accessed 23 July 2010)

As he paced he mulled over his ideas, observations and results of his experiments, trying to make sense of them and making connections between different ideas and types of knowledge. The fresh air, gentle exercise and rhythm of walking helped his thinking processes and it was during these walks that Darwin began to make sense of his personal theories that became the basis of his writings.

Today Neuro-Linguistic Programming (NLP) recognizes that there is a link between mental processes and physical movement:

> Repetitive physical movements and activities involving major muscle groups such as (walking, swimming, biking, playing tennis, etc.), influence our overall state of mind, and thus provide a more general context for our thinking processes.
>
> (www.nlpu.com/Articles/article6.htm accessed 15 February 2010)

What is the reality?

Teachers in the schools that have developed a Darwin Thinking Path have recognized the benefits of a designated Thinking Path for children when working in science, for example:

8.2 *A Darwin Thinking Path*

- linking physical exercise with cognitive processes
- giving children time out and space to think
- showing children that thinking space is valued
- understanding that in order for children to be creative, they need time to engage in critical reflection
- understanding that being outdoors can help to stimulate ideas and encourage children to make connections
- highlighting that thinking is a skill and Personal Capability that can be enhanced and developed. Metacognition therefore emerges in terms of children thinking about their thinking, their feelings and themselves.

The Darwin Thinking Path project has shown how teachers can help children to understand the link between their path and Darwin, modelling their methods on those of a great scientist.

What are the options?

So how can a Darwin Thinking Path be used to support children working 'Beyond the Classroom Boundaries'?

Children need time out: they require time to think about their science; Sternberg (1999: 25) quotes Bethune who, as long ago as 1837, suggested that a creative person can 'store away ideas for future combinations' but what is required is the opportunity and space for an individual to take time from a busy classroom and curriculum to make those combinations or connections in their thinking. As Feasey (2005: 22) suggests a 'thinking environment is characterised by opportunities for the children to have time to engage in critical reflection, so that a gestation period might help children to form their ideas' just as Darwin needed time to think and be creative.

Giving children time out to think about their science, make connections, talk with a fellow learner as they follow 'Darwin's Thinking Path' might seem to be a luxury, but it should be an integral part of science. It is therefore important that children have time to think through ideas and appreciate that, like Darwin, they can take inspiration from nature to help them understand their world. It is also important that children understand that like Darwin, they too can have access to have fresh air, space and time out from the confines of the classroom to think and generate ideas.

Children might tread this Path for a range of purposes, for example, to:

- think through their ideas
- plan investigations
- observe using all the senses
- note changes in the environment
- problem-solve
- clarify ideas and understanding
- consider how their feelings affect the way they generate ideas, put forward viewpoints and share opinions
- note how the physical environment influences their attitude to learning, the cold, heat, wind, and so on.

Setting up the Path

School grounds are a free asset and setting up a Darwin Path does necessarily require any expenditure. It does however does need everyone, both staff and children, to take ownership and to participate in creating and using it. Here we share ideas from a school that have set up their own Darwin Path.

Castleside Primary School Science making the Thinking Path early years friendly

Sarah's 'Darwin Thinking Path' was created around the perimeter of the school grounds but avoiding the Foundation/Key Stage 1 area, since it was thought that the older pupils would be distracted by children in the early years area.

However, the path was designed for use by everyone in the school including those in the Foundation Class and Key Stage 1.

As the Darwin Thinking Path began to be used by children across all year groups it became obvious that it really had not taken into account the needs of the youngest children in the school. Originally the path had 'thinking' words hung around it as waymarkers that could be changed. What she found was that it was difficult to recognize it as a walk, so, cleverly, Sarah decided to purchase plastic feet and hands, and place them along the Thinking Path to signpost for all the pupils but especially children in the early years classes where the path went (feet) and what they might like to touch on the way round (hands).

This led Sarah to consider the other senses and she realized that the same approach could be used to encourage children to use all of their senses except of course taste (for health and safety reasons). She extended the idea by including pairs of eyes, suggesting to children that they pause and observe using the sense of sight and an ear sign to encourage children to stop and listen to, for example, birds or plants rustling in the breeze and of course a nose to smell herbs or flowers.

Then with her Key Stage 1 class they took this a stage further and decided to create Thinking Path stones. These were large stones onto which the children painted inverte-

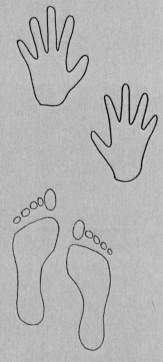

8.3 Thinking feet and hands

brates that were found around the school grounds, such as wood-lice, beetles, worms, butterflies and ladybirds. They looked really brilliant and the children placed them along the Thinking Path to encourage children to stop and look for these animals. A tip from Sarah though, make sure that you use the right kind of varnish for the stones, otherwise the rain simply makes the paint run!

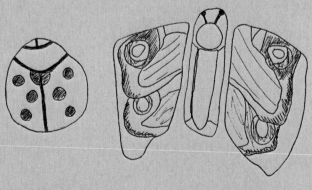

8.4 Thinking Path stones

We must not forget though that taking time out to 'cogitate', 'muse' and enjoy and be inspired by nature is also important to children's spiritual development and we should aim to offer children opportunities to experience awe and wonder of the world around them.

What will you do?

You could begin by sharing the idea of Darwin's Thinking Path with staff, children, school governors and parents by showing them a clip from the film *Creation* where Darwin is walking his path. Some schools have had a special school assembly held outside to launch the idea of Darwin's Thinking Path and invitations have been sent to friends of the school. Celebrate the launch of the path with a Thinking Path special event.

In the meantime in order to set up the path there are a number of questions that need to be considered, such as:

- Where in your school grounds could you have a Darwin Thinking Path?
- How can you ensure that staff understand the philosophy behind the Thinking Path?
- How will staff make sure that the children are given sufficient time for the path, so that they can observe, reflect, discuss, think, share ideas and collect information (e.g. data, photographs)?
- When will you take staff round the Thinking Path so that they can decide how to use it with their own class?
- How will you and your staff explain to children the purpose of the Thinking Path?
- How would you encourage staff and children to take ownership of the path?
- What 'waymarkers' could be placed around the Thinking Path to encourage and support its use?
- How would you maintain the Thinking Path encouraging children and teachers to use it on a regular basis?

Practical ideas for using Darwin's Thinking Path

Darwin walked his Thinking Path on a regular basis and while we do not expect children to walk the path for an hour every day, we would expect them to use the path regularly. Use of the Thinking Path will vary but might include children walking the Thinking Path:

- in their own time during playtimes
- when they need some time out to reflect on a problem in a science lesson
- at regular intervals, for example, once a week, fortnightly or monthly, to note changes in the environment
- as part of a science lesson, e.g. on habitats, sound and light.

When children use the Thinking Path they could be encouraged to observe and think about any of the areas suggested here.

What have you noticed?

- What have you noticed on your path today that you have not seen before?

- What is the light like today? How does it affect the Darwin Thinking Path? How does it make you feel?
- How many new plants can you see? How many invertebrates can you spot today?
- Talk into the Easispeak microphone and describe what you can see, hear, smell and touch while you are walking the path. Explain how you feel.
- Make a note of what is new on the Darwin Thinking Path board.
- Stop and take a photograph, store or print out and annotate it with information, and display it somewhere along the Thinking Path.
- Stop and photograph the same spot every two weeks to record changes over a school year.

Similarities and differences

- What things are the same along the Darwin Path?
- What things are different along the Darwin Path?
- How does the Darwin Path change during the day?
- What do you think the Darwin Path will look like in autumn, winter, spring and summer?
- How many plants are the same?
- How many different places can you find where animals live?

Plants

- How many different plants are there along the Thinking Path?
- How can you record the different plants?
- What are the plants called? How will you find out?
- When you have found out the plant's name design and make a plant label, make sure that it can withstand the weather.

Trees

- Adopt a tree on the Darwin Path, perhaps it is your favourite tree.
- Why do you like your tree? How does it make you feel?
- Take photographs around of your tree in the autumn, winter, spring and summer to show how it changes.
- What can you find out about your tree, for example: what do the leaves look like? Why don't you get some paper and do a bark rubbing. How can you find out how far it is around the trunk of the tree?
- Do any animals live on your tree? Which ones? Do any animals visit your tree? How do you know?

Seeds

- How many different kinds of seed can you find?
- Which is the biggest seed and smallest you can find?
- How are the seeds dispersed?
- Which plant did the seeds come from?
- How did they get from the plant to where you found them?

Animals

- How many different animals live along the Thinking Path?
- Observe and sketch the animals that you can see, e.g. a snail.
- Find out about each animal and its habitat.
- Why are the animals found there? Are the animals resident, or visitors?

Yourself

- What are you thinking about today when you walk along the Thinking Path?
- What things in science are you thinking about?
- How does walking along the Thinking Path help you?
- How are your feelings affecting your thinking about science?
- Who are you with, or are you on your own?
- How does that help you think about your science thinking?
- How does talking to a friend help you work things out in science?
- How is the weather affecting you today – is it cold, hot, windy?
- How does the weather affect your thinking?
- What do you think about what you have been asked to do today?
- By your teacher, your group, your partner?
- Will you be someone who thinks interesting things in science today?
- Are you thinking, yes, I can think interesting things today?

References

Feasey, R. (2005) *Creative Science: Achieving the WOW Factor with 5–11 year olds.* London: David Fulton Publishers.

Sternberg, R.J. (1999) *Handbook of Creativity.* Cambridge: Cambridge University Press.

www.bbc.co.uk/history/historic_figures/darwin_charles.shtml

www.greatplanthunt.org

www.nlpu.com/Articles/article6.htm

9 Science Role-play Sheds 'Beyond the Classroom Boundaries'

What is the goal?

Teachers and children often find it frustrating when equipment relating to outdoor activity has to be brought back indoors or stored elsewhere. Investing in a Science Role-play Shed will offer significant payback!

The idea of a Science Role-play Shed is to create a self-contained area which can be changed and themed according to the curriculum and the interests of the children. It offers role-play opportunities with a science focus and develops children's personal skills and capabilities.

The real benefit is that the whole shed becomes the role-play area and those resources that children take out can be put back inside the shed, the door closed and padlocked, and each morning opened again ready for children to embark on further exciting science-based experiences.

9.1 Science Role-play Shed

What can the Science Role-play Shed offer?

- More space than the classroom.
- Use of and immediate access to the external environment.
- Inside and outdoor spaces that are juxtaposed.
- Internal surfaces for children to use.

- Opportunities to redesign the inside and outside of the shed relating to the topic.
- Doors for hanging resources and activities.
- Self-contained area for creative science activities and role play.
- Workshop 'apprentice' area where adults and children can work alongside each other.
- The ability to close and store the science role play until the next day.

While we would accept that role and free play are important elements of the early years experience, there are areas of science where children can learn a great deal from adults that will allow them to develop new expertise in their role play, their science curriculum activities and at home.

What's the reality?

Children are used to learning about people who help others, often in relation to health and caring. Teachers tell us that even though children have access to different people such as the school nurse, school cook, vet, optician and fire service, the scientific aspects of these jobs are rarely pointed out to the children.

People with different areas of expertise require an array of different scientific knowledge, skills and personal qualities. For example, the school cook needs to know about food groups, balanced diets, changes in materials, reversible and irreversible changes, and heath and safety in the kitchen. The school nurse must know about the skeleton, body organs, keeping healthy through exercise and nutrition, personal hygiene, and the effect of drugs including medicines as well as the roles of people in the health system: for example, doctor, nurse, physiotherapist, radiologist. Both the cook and the nurse need to know how to respond and communicate effectively within their peer groups and with others – for example, parents, children, specialists, trainers and so on – and to be willing to take instruction and manage their workloads in order to reach deadlines.

There are some occupations where it is easier to identify how science is used, whereas in others it is less obvious but equally important. While some of the occupations, such as those of the cook and school nurse are probably more appropriate for indoors activities, there are others that can very easily be used as the basis for the science shed. A summary of these can be found in Table 9.1.

What are the options?

First, either purchase or reuse an existing shed, making sure that it is safe for children to use. Ensure that there are no rough surfaces where children might get wood splinters or grazes, and do make sure that the doors can be opened from the inside as well as fastened back securely so that they cannot be knocked or blown shut. When setting up different experiences for children, consider how to offer children the experience of learning new skills and knowledge from a range of adults with science expertise and what opportunities the experiences will offer for the development of Personal Capabilities.

Table 9.1 Occupations and science

Job	Science	Links to science curriculum	Essential Personal Capabilities
Fire service	• Flammability • Properties of materials • Effects of fire and smoke on human body • Substances and extinguishing fire • The fire triangle	• How fire starts • What fire needs • Materials in the home • The human body	• To be aware of oneself and others • To accept and deal positively with risk • To exhibit self-control and good communication skills
Garage mechanic	• Forces • Movement • Electrical systems • Materials and their properties • Liquids • Health and safety	• Making things move • Gears • Pulleys and levers • Lubricants • Friction • Materials (e.g. transparency, flexible, hard, scratch-resistant) • Working safely	• To explore and resolve problems • To apply knowledge from one context to another • To organize and plan how to go about a task
Vet	• Animal groups • Animal physiology and life cycles • Animal habitats • Animal illnesses • Keeping animals healthy and caring for animals	• Different animal groups (e.g. mammals, birds, reptiles, invertebrates) • Animal parts and skeletons • Animal life cycles • Caring for pets • Animal habitats and homes	• To be empathic and sensitive to others • To explore and resolve problems • To analyse situations • To keep track and monitor progress

(Continued Overleaf)

Table 9.1 Continued

Job	Science	Links to science curriculum	Essential Personal Capabilities
Builder	• Health and safety • Properties and uses of materials • Forces • Movement	• Properties of materials • Uses of materials • Forces to make things move and change shape • Electricity, circuits, safety • Health and safety on a building site – what people wear, how they work	• To organize and plan how to go about a task • To be aware of and limit risk • To meet deadlines, overcoming failures or difficulties • To persevere in the face of adversity
Gardener	• Plants – classification, habitats • Plant life cycles • Plant propagation, cultivation • Rocks and soil types • Animals – pests, pollination • Forces • Pest control • Effects of seasonal changes and weather	• Plants, parts of a plant, plant identification • Animal needs • Plant life cycles • Growing plants • Animals and plants	• To be creative and imaginative • To apply knowledge from one context to another • To evaluate progress and influence change • To organize and plan
Conservationist	• Animals • Plants • Climate • Human effects on the environment	• Different animal and plant groups • Habitats • What plants and animals need to live	• To be self-aware • To influence and persuade others • To take responsibility • To show perseverance and commitment

	• Conserving the environment • Stewardship	• How we can look after our own environment • Making positive changes to our own environment • Understanding responsibility to our environment – stewardship	• To communicate to different audiences
Archaeologist	• Rocks • Soils • Animals and plants • Habitats and environments • Body parts • How the Earth has changed • Collecting, storing and using forensic evidence	• Rocks • Soils • Animal and plant classification • Collecting and using evidence • Rock classification	• To analyse evidence • To explore different possibilities • To be organized • To reach conclusions based on evidence

Children as apprentices

The idea of the science shed as a role-play area fits well with early years provision, as does the idea that children could take on the role of apprentice under the tuition of an adult or older child in the school. As apprentices the children will have authentic experiences where they learn a skill, new scientific knowledge and develop Personal Capabilities related to the science topic in the shed. This will require the school to explore or make use of existing links with the local community so that a 'bank' of people with varying expertise can be asked to work with the children at different times during the school year. The potential for this is limited only by the imagination of the teacher and children, and the number and kind of experts that are able to work with the children. The following example provides a context which brings together role play, occupations and experts from the community in creative and innovative approaches to using areas 'Beyond the Classroom Boundaries'.

Wheeler's Garage Science Role-play Shed

A garage is a common role play for early years children and it is also an excellent setting for children exploring aspects of science and developing Personal Capabilities.

Making experiences more authentic for children is important and can be achieved by visiting a local garage and inviting a garage mechanic into school to work with the children. While there is a world of

9.2 Wheeler's Garage Science Role-play Shed

difference between a mechanic working on a car for a customer and on toy cars there is a lot of overlap in terms of science. When developing the Mechanics Garage Role-play Shed think about including the following:

- overalls
- tools (e.g. spanners, bicycle pump, screwdrivers, pliers)
- large blocks that toy cars can be placed on so children can 'safely' work underneath
- car wash area, where children can wash cars, bicycles, and so on using everyday washing up liquid, hose if available, and make choices about whether to use a chamois leather sponge, rags
- real tyres (e.g. bicycle, car)
- hazard and safety signs
- funnels
- empty containers such as lemonade bottles, cartons with labels (e.g. oil).

Approved members of the local community who are or have been mechanics, such as retired grandparents and local business people, are superb support for children if they can be engaged in building their own model cars or buggies. Children could use different construction kits to make a vehicle that moves, or they could construct a go-kart made from boxes, wheels, string and so on. The children could be engaged in the design process on paper or be offered a range of materials which the mechanic can help children to use to make their kart. What is important is the interaction between the children and the mechanic, which needs to be planned in terms of helping to develop Personal Capabilities and the science. It is of course appropriate to provide the adult with key areas of science that you want children to engage with; you might offer a set of simple statements such as:

- a push or pull force can make things move
- a force can slow down or stop something moving
- push and pull forces can change the shape of things
- the bigger the push or pull force the further something moves and the more an object changes shape
- forces can make a moving object change direction.

This not only will help the adult appreciate the area of science the children need to access but also provide him or her with basic vocabulary when talking with children.

Activity

Over the period of using the Mechanics Garage Role-play Shed offer children different challenges to engage with to provide problem-solving scenarios. Children might arrive one day to find a new job sheet, or a letter asking the children to solve a problem or create a new vehicle. You could set out a range of construction kits such as Lego, Duplo, K'NEX and Mobilo; even better would be to put out components from very large constructions kits, to make a vehicle, which might have to:

- carry an object of a certain size
- move over sand
- move over rough ground.

To help the children work through their problem they could be encouraged to begin by coming up with different options to consider. Encourage children to use strategies such as 'think–pair–share' which will stimulate them to think about the problem, pair up with a partner or adult helper and share their ideas. They could compile their ideas, using diagrams if desired. This simple approach offers the children some support in sharing ideas and helps them begin to verbalize their thinking. Even the very act of sharing their idea can be a difficult thing, especially for younger children who may feel apprehensive to put their idea forward to a larger group.

What will you do?

Of course the first thing is to decide whether or not you have an existing shed that could be used: for example, at Gainford Primary School staff teaching early years realized that they had a shed which had in their own words become a 'dumping ground'. Suddenly they had a great outdoors Science Role-play Shed which the children love.

If you do find you need to purchase a shed think carefully about the safety issues indicated earlier in the chapter and how to secure the shed at the end of the day.

Once you have your shed the only limit is your imagination! Think about which topics would lend themselves to the Science Role-play Shed idea and what areas of scientific understanding and potential for developing Personal Capabilities the shed has to offer. Equally important is to consider which year groups will have access to the Science Role-play Shed, remembering that the aim is to offer these experiences across the 3–7 age range. In the final section of this chapter we offer some practical suggestions for topics that might form the role-play element of your science shed.

Practical ideas for creating Science Role-play Sheds

The Archaeologist's Dig Science Role-play Shed

This is often an indoor activity where children are given jigsaws of dinosaurs, or items in a sand tray such as dinosaur pieces or artefacts. How much more interesting for children to be able to set up an Archaeologist's Dig 'Beyond the Classroom Boundaries' and to include a workshop where mess on the floor is of no consequence, with all kinds of exciting activities and equipment.

Imagine the 'Science Shed' and surrounding area becomes an archaeologist's dig where the children can go and put on their archaeologist's coat (white shirt just larger than the child's own size)

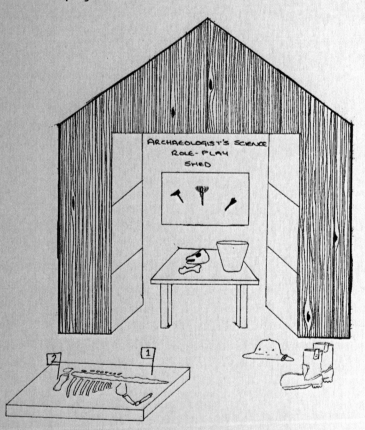

9.3 Archaeologist's Science Role-play Shed

and dig the area to find fossils, bones, coins, tracks and bracelets. Once the evidence is collected they then go into their laboratory and use the evidence to determine to what or whom the items belonged.

Outside the laboratory are areas that are coned off, with signs that indicate that only archaeologists with identity passes can access the area. Once inside sand trays and soil trays might contain rocks with fossils, old coins that are covered with hardened mud so that they have to be brushed clean, and pieces of old bracelets (old jewellery). Under supervision children could make casts of tracks left in mud that has dried or take photographs that can be printed out in the classroom and taken to the laboratory and hung up.

Inside the laboratory might be fossils, bone sets, tracks and plant prints that the children have to work out what it was and explain the evidence for their thinking. Children might be challenged to complete drawings of fossils or complete a whole skeleton; they could be asked to fill in written reports or give a verbal report using an Easispeak microphone to record observations and conclusions.

There are many different science skills that can be developed in this type of role play, from seeking and finding evidence to recording what they find and drawing conclusions. At some point children could communicate to the rest of the class what their finds were for that day, which, of course, would be different from what has been found previously.

Access to an adult who has expertise relating to rocks and soil or to archaeology can make a dramatic difference to the quality of this science role-play scenario. One of the best places to locate such experts will be your local university, which will have both tutors and students who might be willing to join your children to share expertise and techniques with them.

Activity
Put out a large sand tray with tape around it, making it look like a real archaeological dig, renewing it for each new group that is scheduled to work in the dig. Give the children their shirt, shorts and builder's-type helmet to wear and tools such as trays, paintbrushes, plastic trowels and tell them that they have to find the animal bones. These could be dinosaur bones created from paper maché (or cleaned animal bones from the butcher) which when found and put together could make a small prehistoric bird, fish or mammal. The children should work in teams and have to work just as an archaeologist does, by laying out the bones, putting a label on them with string, and so on.

Working in a team is sometimes taken for granted, but remember that during this activity we should endeavour to support the children in their team roles as much as possible. Team role badges are a useful idea in this activity. Consider a team leader who oversees the group as a whole, a team digger who unearths the finds, a team investigator who must clean up the find and describe it to others, a team reporter who draws or takes photos of what was found.

Space Rocket

It is not difficult to imagine how the 'Science Shed' could metamorphose into a 'Space Rocket'. Out go many of the items from its previous role-play scenario and in come posters and photographs of planets, galaxies, the Moon, life size astronauts, control panels, microphones and walkie talkies for communication with Earth. Inside the Space Rocket on the walls and ceilings might be glow-in-the-dark stars and a range of equipment for astronauts like control panels and portals showing galaxies. Telescopes of course are a must so that children can see a

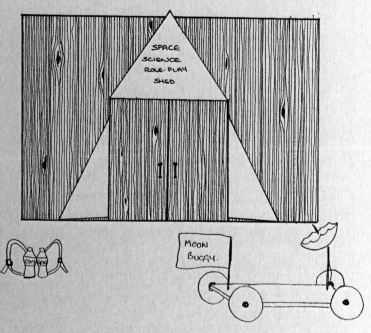

9.4 *Space Rocket Science Role-play Shed*

distance but with the safety warning that they must not look at the Sun.

Astronaut suits are common in role-play clothing and of course all good rockets have planet rover buggies for when they land and go out to explore the planet. The area outside the Space Rocket might be given a temporary, uneven surface using sand and play rocks, and craters to make it more authentic. Litter pickers are great for astronauts to pick up planet rocks, and children can make their own oxygen apparatus using plastic bottles. Of course, the Moon surface would not be complete without the addition of a Moon Buggy, changing a car to something special with the addition of silver foil, antennae and whatever you and the children can think of to make the experience more authentic.

There are many people who can offer expertise to children and again a number of these will be from university departments such as astronomy or astrophysics. In particular, University Outreach Teams will be willing to work with children sharing an understanding of space and will also help them to use items such as telescopes.

Activity

Bring the children together and introduce the idea that the Role-play Space Shed needs a space shuttle, so that it can carry some astronauts to land on and explore a nearby planet. Challenge the children to use materials that have been placed outside, to design and make the space shuttle, including the inside which might require seats, control panels, somewhere for the astronauts to sleep, and so on. You could utilize the outdoor climbing frame if you have one, or use very large cardboard boxes, and also offer children:

- silver foil
- carpet underlay
- plastic bottles
- plastic crates
- sleeping bags
- old curtains.

The creative approaches that ensue from this activity are multiple. Talk with the children about what it means to be creative, ask them to take some time out to think about how they can make the best possible rocket so that it can do its job well, and be a stimulating place for the astronauts to work in. Allow children to use chalks on the ground, whiteboards or large pieces of paper to draw out some ideas, although some may want to get straight into making and constructing. Creativity at this age is very much about excitement with a purpose, so it would be useful to stop every so often to ensure that the purpose of the rocket is being maintained.

Conservationist's Science Role-play Shed

This is a great idea for young children who become very keen to help to plan and look after the natural world in their school grounds. What is even better is there are so many organizations today willing to come in and work with children to teach them about their environment and how to look after it, as well as plan for future sustainability. If your school is an Eco-School, this will fit in well with your school's mission which will include the need for children to work within the school and in the consultation with the wider community on chosen environmental themes such as biodiversity and school grounds. Both of these fit perfectly with the idea of children working 'Beyond the Classroom Boundaries' and the Science Shed becoming the 'Conservationist's Den', 'Eco-Den' or to adapt titles from popular TV programmes that children view such as 'Bug Watch Den' or 'Spring Watch Den'.

The children could engage in collecting and keeping invertebrates, photographing them, using computer microscopes and identification charts. From the Conservationist's Shed they could birdwatch, make collections of natural objects and, even better, if they had access to wi-fi and webcam to a bird box in the school grounds. Adults working with the children could help them to create a wild area with insect boxes, plants in containers and log piles to attract animals, even if it is only invertebrates such as woodlice.

Inside the Science Role-play Shed there could be a whiteboard with words and or pictures of the jobs for the week or the day. They might include:

- weeding vegetable or flower beds
- sowing flower and vegetable

- potting on plants
- planting out flower and vegetable beds
- sorting seeds that have become mixed up and are in the wrong containers
- carrying out a 'bug hunt' and logging which invertebrates they find in the school grounds and where
- using binoculars to birdwatch.

9.5 Conservationist's Science Role-play Shed

The 'Conservationist's Science Role-play Shed' can offer children a variety of different experiences but will be enhanced by inviting 'experts' to work with children, preferably in the early days of children's access to the shed. Local gardeners (and this could be members of the school's governing body or even the caretaker), as well as people from environmental groups, could be asked to work with children to teach them a range of skills that they can apply when they 'work' in the shed. Interestingly, where children have been given access to these kinds of expert in project schools not only has this enhanced their Personal Capabilities and allowed them to engage in a range of activities in school but also many children use their newly acquired understanding and skills at home, sharing them with family.

Activity

Show children either pictures of animal homes or bring some in: for example, a bird box or a lacewing home and discuss why we put these in our gardens. Then ask the children to design and make their own while they are in the Conservationist's Role-play Shed. If you can arrange for an adult with woodworking skills to work with the children it would provide an authentic context for this activity. This would offer children the opportunity to work as apprentices to the adult, and learn new skills to create something that could be placed in the school grounds, which would motivate children to check on the home on a regular basis across the year.

Activities such as designing and making animal homes as in Illustrations 9.6 and 9.7 encourages creativity and ensures that children have to think about fitness for purpose, which demands that children reflect on what they have produced.

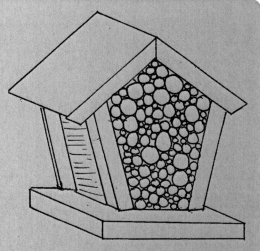

9.6 *Lacewing home*

9.7 *Bird box*

The Garden Centre Science Role-play Shed

The garden centre is a popular role-play area in early years, and would make a great Science Role-play Shed. The usual way to set up a garden centre is to take the children on a visit to a local garden centre so that they can see what happens there, how things are set out and what they sell. If possible, do ask if someone at the centre can teach children how to plant seeds and pot on plants. Even better would be to arrange for someone from the garden centre, or a parent to work with the children in the Garden Centre Science Role-play Shed on a regular basis, so that they can learn various gardening techniques, about seed and plant growth and learn to converse with an adult using scientific language related to plants.

9.8 *The Garden Centre Science Role-play Shed*

When setting up this role-play shed do think about sending out a letter to parents and local garden centres asking for donations of clean and safe resources; don't be shy, do offer them a list, it is easier for people to look at a list to see what they have, for example:

artificial flowers	garden ornaments	posters of	toy
bags containers	(e.g. gnomes)	plants, plant	lawnmower
buckets	measuring equipment	life cycles	toy
buckets canes	phones	seed collection	wheelbarrow
children's fork	plant labels	seed packets	watering
and trowel	plant pots	seed trays	cans
cloches	plants real and	sieves	
compost	plastic	till and money	

There are many activities that children could role play in the garden centre including:

- creating a new garden using containers such as large pots, old tyres, unusual containers such as a teapot, a shoe
- giving planting demonstrations
- making seed packets with instructions
- measuring plants and keeping a plant diary in the shape of a plant, flower or seed
- planting up hanging baskets
- planting seeds
- potting on plants
- pricing up plants for sale
- selling plants and garden equipment to customers
- sieving soil
- sorting and ordering plant pots
- sorting seeds into the correct seed packet, e.g. acorns, runner beans, broad beans, peas, sunflowers, sycamore, beech, horse chestnut, avocado
- using magnifiers to look at plant parts and seeds
- watering and tending plants
- writing plant labels for plant pots.

Do not forget that the children could engage in growing plants to sell to parents: for example, having a garden centre coffee morning where the children sell their plants, seeds and so on to raise funds. Do make sure that children have created labels, price tags and signs such as seeds, garden plants, indoor plants, herbs, for sale, open and closed, and so on.

The Personal Capability in this activity could focus on good communication. Any good garden centre should provide its visitors with clear signs and symbols so that people know where to go, which plants may be hazardous and information about how to care for particular species. Such communications can be transmitted in different ways: drawings, symbols, clipart, Easispeak microphones, photographs and so on.

Activity – Jack's Mixed-up Bag of Bean Seeds
Read any of the *Jack and the Beanstalk* stories: for example, *Jaspers Beanstalk, Jim and the Giant Beanstalk* and engage children in a discussion about growing beans. You might wish to change the story where Jack has been given a bag of beans but when he opens the bag he finds that they are not the same, which raises questions such as:

- What kind of beans are they?
- What kind of bean plant will they grow into?
- Will some grow more quickly than others?
- Which one will be the tallest?

Give children a set of mixed-up bean seeds of different varieties to sort, including mung beans, broad beans, runner beans, lima beans and French beans. Once the children have sorted the seeds into the correct seed packets, then they have to follow the instructions to plant them. Give children a small garden area, large trough or plant pots to plant their seeds and challenge them to think about creating a plant label and perhaps a picture of what the bean plant might look like.

Jack's Bean Garden

Kay at Wheatlands Primary supported children in creating Jack's Bean Garden, where children grew different kinds of beans in different plant pots. The children created a photographic diary of the plants and also created similarities and differences cards for other children to use displayed around Jack's Bean Garden. Persevering with growing plants can be frustrating for children, who are more used to instant results,

9.9 Jack's Bean Garden

rather than having to wait for a seed to germinate and plant to grow. These kinds of activity in science are important for developing not only children's concepts of life cycles and change but also the self-management capability of tenacity. Keeping track of the beans, logging their development, reviewing their progress and considering what should be done next are some of the skills that can ensue from this activity. Talk with the children about how they will monitor what is going on, when they will log the growth, how and by whom. Encourage them to control things that might not be going their way by asking them what they might do if they think the beans are not growing as well as they had predicted.

The Theatre Science Role-play Shed

Children just love theatres, role-playing and being part of the audience, so the Theatre Science Role-play Shed fits in not only with what children in the early years enjoy but it also has brilliant science potential, particularly if children are engaged in shadow puppet work.

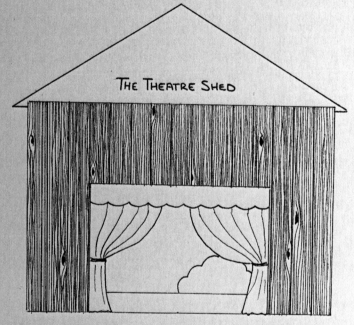

9.10 *The Theatre Shed*

As with all of the other science role-play sheds please think about linking up with an 'expert' in this area who can help children to:

- make puppets
- access stories
- design sets
- teach puppetry
- help children to understand how shadow puppets work in terms of light and shadow.

Shadow hand puppets are of course the cheapest and easiest, allowing children to explore and be creative using their own hands. There are lots of hand shadow pictures to be found on the Internet; do print some off for children to look at how hand shadows are made. Do explore materials such as paper doilies as the basis for puppets; these are easy to make and look spectacular because of the patterns that they can make, as shown in Illustration 9.11.

There are a number of books for young children which lend themselves to being used as a starting point for activities in the Science Theatre Role-play Shed: for example, challenging children to retell Mick Inkpen's story *Kipper's Monster* by making their own shadow puppets using card.

One of the great things about puppets is that most cultures have puppets and stories that go with them so science can take on an international perspective in terms of children being told stories from different cultures and using the Theatre Shed to recreate the story as a play, having made the characters themselves.

9.11 Paper Doily Shadow Puppet

Activity

Working in pairs children research the life cycle of an animal and use this knowledge to create their own shadow puppets and play to tell the life cycle of their animal (e.g. a spider, a butterfly, a chicken or a frog). With the shadow puppet play the children have to provide the narrative to explain to the audience the life cycle.

Developing the narrative for this type of work is challenging for children of this age; therefore, it may be useful to compile it together in groups of five to six children supported by an adult or older child, and to use a big book or floor book. Sharing ideas, taking turns, listening to others, reaching agreements and co-operating in groups of this size are vital Personal Capabilities that you can encourage.

CHAPTER 10

Science Story Sacks 'Beyond the Classroom Boundaries'

What is the goal?

Feasey (2006: 8–10) wrote that 'stories are the mainstay of primary education and children of all ages love them' but why restrict them to science in the classroom? Why not take stories outside, and even better, why not let the children take them outside and read them as the starting point for exploring science 'Beyond the Classroom Boundaries?' The idea for this chapter came from Kay Coverdale, a teacher in Redcar, who had been on a course that suggested that Science Story Sacks containing a book and science equipment could provide a starting point for science lessons. Most teachers are familiar with the idea of story sacks, since these have been used in schools for a number of years, but not Science Story Sacks. Kay's response was to take the idea and then be creative in using the concept to suit her school, children and resources. In this chapter we share and extend Kay's innovative idea of developing Science Story Sacks so that they can be used to explore science 'Beyond the Classroom Boundaries'.

What are Outdoors Science Story Sacks?

Quite simply these are story sacks (bags) which contain a story book that has either a science context or potential for exploiting science outdoors and equipment, which children can use outdoors to explore their environment. A very simple concept which aims to:

- use a familiar approach to learning as the vehicle for science 'Beyond the Classroom Boundaries'
- show how stories can be used as a 'hook' for science
- illustrate how stories can help children to understand contexts for science
- show applications of science
- help to develop a range of Personal Capabilities.

What is the reality?

The reality is that while stories are a familiar context for learning in many areas of the curriculum, in the primary years, stories are used less frequently to engage and motivate children in science. Very early years teachers use stories in science: for example, *Jack and the Beanstalk*, when growing seeds; and *The Three Little Pigs* for teaching materials. However the practice is not widespread, particularly as children move from the Foundation years through the school. The range of stories is also limited and often teachers do not recognize fully the science that can be developed from them, and even more true in the case of Personal Capabilities.

The idea of an Outdoors Science Story Sack is innovative and offers many exciting opportunities for teachers and children. Making it into a reality just requires lots of imagination, enthusiasm and a little determination, but as Kay found out, it is well worth the effort.

What are the options?

Kay's Story Sacks

'I am really proud of my Outdoor Science Story Sacks, a lot of it came from a stories and poems course I attended. I could see where science could come from but felt that if they were only slotted into lessons use would be limited and I wanted the story sacks to be accessed at play and lunch times. So I decided to make Outdoors Story Sacks using colourful bags with lots of science resources inside linked to the book in each sack. Initially the idea was that the younger children could take an Outdoors Story Sack and go off, read the book and do the activities. So I used some books from the school library and purchased some new ones, a wide range including some picture books and very simple language books.

I have now extended the idea so that older children can work with younger children as an 'end of day' activity, where Year 4 children pair up with Foundation children and also those in Key Stage 1 and read the story and use equipment in sacks to explore science in the school grounds.

To support this, the Science Story Sacks now have:

● an instruction card for older child for what to do

● a fact card to extend older child's science subject knowledge

● a suggestion card for what they can do with equipment although they could do what they want

With the Outdoors Story Sacks the children have the responsibility: each sack has a label indicating what is in the sack and therefore they have to make sure that everything is put back. The children become the experts and they share their knowledge with others and choose which sack they would like to use.

> When I first took them into staff room, all the staff were looking at them and wanted them, they tucked into the bags and would you believe they were playing with contents, the bags had staff so excited.'

Sound Story Sack – Quiet! by Paul Bright

In this section we look in more detail at the contents of one of Kay's Science Story Sack and her rationale for including each of the items.

Inside the bag are the following items:

The book Quiet! by Paul Bright

Quiet! is a story with lots of potential as a starting point for science 'Beyond the Classroom Boundaries', since it is about a father lion who is trying to keep the animals quiet so that his lion cub can sleep. So, Kay made the focus of the items in the Outdoors Story Sack the topic of sound.

Science equipment

In terms of science equipment the aim is to fill the Science Story Sack with items that will motivate children to explore their outdoor environment. An important issue here is that where possible the equipment should be familiar to the children, since

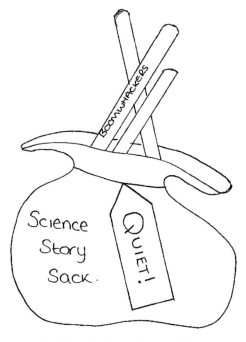

10.1 Outdoors Science Story Sack – Quiet!

the sacks are designed so that they can be used, where appropriate, without adult supervision and intervention. So for the science from this story, Kay put Boomwhackers, which are plastic tubes of different lengths into the Story Sack. The accompanying activity card suggested that children could use them to tap different objects and materials around the school grounds to see what sounds they made.

Activity card

To help children use the equipment, each sack has an activity card, in the 'Quiet!' Sack the card encourages children to make quiet and loud sounds, sounds that will not wake the lion cub and those that would. Depending on the age and ability of the children the cards could offer written prompts or photographs showing other children using the Boomwhackers.

Background science knowledge card

Kay's story sacks have been designed so that older children can read the story to younger children and support them in exploring the ideas for activities. She felt that the older children might need some prompts relating to the subject knowledge to share with the child they were working with.

Puppets

A number of the stories that Kay has chosen to use for the Science Story Sacks have books with characters for which there are puppets, so in this sack Kay placed a lion and a lion cub. A lovely idea and one which allows either the reader to tell the story using the puppets or for the child listening to then use the puppets to role play the story or use them when exploring with the Boomwhackers.

A unique aspect of these sacks was the way in which they encouraged discussion and recognition of particular skills and capabilities, such as creativity, problem solving, teamwork, self-management and communication. The purpose of these activities is to develop knowledge, understanding and language of Personal Capabilities in a relevant way. Activity cards provide stimulus for adult helpers or peers to highlight the skills characters are using, and to engage children in thinking about these in relation to themselves. Such activities are generic tasks that can take place outdoors or indoors, at the same time as the science activity or during the same day or week.

What will you do?

Interestingly, Kay made her Story Sacks and then showed them to staff at her school who immediately began to explore them and decide how they could be used. So perhaps the best approach is to have a go with the children in your class, allowing them to perhaps take them outside at play- and lunchtimes, since often when other children see them demand grows!

Of course, there is the issue of putting the sacks up first; there are a number of options:

- purchase drawstring PE bags
- invite a group of parents to make canvas bags
- use a staff meeting to brainstorm ideas for stories and contents
- fund-raise with a science story marathon read to purchase puppets and equipment
- create a list for each book and engage older pupils to put together the Story Sacks
- join forces with the literacy co-ordinator.

Once the Outdoors Science Story Sacks have been created do celebrate by launching their use in an assembly, or perhaps allowing different classes to use them on certain days of the week to introduce them gradually to staff. Make sure that parents know about them, some schools may want to allow the Outdoors Science Story Sacks to be taken home. Do however resist the temptation for them to become indoors sacks. Their purpose should be to encourage children to explore science 'Beyond the Classroom Boundaries'.

Practical ideas for Science Story Sacks for 'Beyond the Classroom Boundaries'

This section provides some ideas for Story Sacks. For some the books lend themselves to a science activity and a Personal Capability activity. Start with these books and add your own selection.

The Snail and the Whale by Julia Donaldson and Axel Scheffler

- Fact cards about sea and land snails
- Fact cards about whales
- Magnifying lens
- Piece of stiff transparent plastic or transparent plastic containers
- Simple identification key for local snails
- Camera

Activity Card 1

- Go around the school, how many snails can you find?
- Where do snails live in the school grounds? Take a photograph.
- Use the magnifying lens to look carefully at a snail.
- Put the snail on the plastic and look underneath to see how it moves.

10.2 Outdoors Science Story Sack – The Snail and the Whale

Activity card 2 (Personal Capability focus – Creativity):
In this story the snail is ambitious; he has a dream and is willing to take risks to be successful. He is also imaginative in the ways he considers he can achieve his goals and also in the ways he solves the whale's problem. Such characteristics underpin creative development; therefore, bringing out these elements from the story are useful in helping children understand how they too may be creative scientists. Pose different scenarios for them to consider creatively:

- Ask them to consider how a snail could get from their house to their school.
- What would we do if an animal arrived in the school grounds and was in trouble?
- How could they make their way to 'X'? (Define a place of their choosing that they feel is far, far away, or somewhere they dream of going to.)
- What are their ambitions and how do they think they will achieve them?

Aaaarrgghh, Spider! by Lydia Monks

- Fact cards about spiders
- Simple identification key for local spiders
- Magnifying lens
- Plastic spiders in case the children cannot find any live ones
- Spider web board

Activity Card 1

- Where can you find spiders in the school grounds?

- What do spiders' webs look like? Draw one.

- Use the spider catcher to catch a spider.

- Put the spider in the magnifying pot to see what it looks like.

- What parts of the spider can you see?

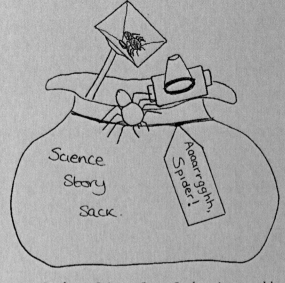

10.3 *Outdoors Science Story Sack* – Aaaarrgghh, Spider

Activity Card 2 (Personal Capability focus: Self-awareness)

In this story we learn about the uniqueness of spiders and the characteristics that make them special and the things they are good at. Children often find it difficult to pinpoint for themselves the things they are good at, so developing a strong sense of awareness can build confidence and also raise self-esteem.

- Tell your friend about four things the spider thought he was good at and that made him different from the people in the story.

- Tell your friend what you are good at and make you different from the people in your class. Do a think pair share with this activity, or if the children find it difficult use Nominate Someone Who, where peers provide the responses based on their judgement of their classmate.

- How can you show people what you are good at more often? How can you make people understand you better? Ask the children to set themselves a target to demonstrate at least one thing they are good at each day.

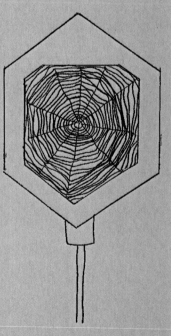

10.4 *Spider web board*

The Treasure Hunt by Nick Butterworth

In the Science Story Sack should be the Treasure Hunt instructions or map.

- The clues for the children to follow could be oral clues on a series of talk Postcards it placed around the grounds, or a set of written clues linked to areas of science; for example, plants, different materials or a force.

- Puppets – the animals from the story, badger, owl, squirrel, rabbit, hedgehog, mole, fox.

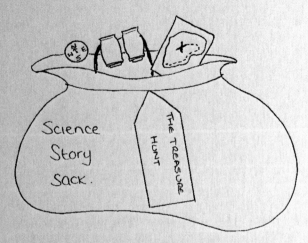

10.5 Outdoors Science Story Sack – The Treasure Hunt

Activity Card 1

- A set of instructions for the Treasure Hunt. The older child might have put the Treasure Hunt cards out before reading the story to the children. The set of cards might be photographs of things in the school grounds: for example, a tree, a bush that has a distinct leaf, items where the children have to match the picture to something in their environment.

Activity Card 2 (Personal Capability focus: Communication)

In this story Percy leaves messages for the animals in various forms, in words and messages. The cow also tries to communicate with them in her own way. Good communication comes in different forms and it is worth encouraging the children to think about the different ways they receive information from other people. Focus on instruction-giving:

- Set a trail around the school in the same way that Percy did, using various forms of communication: for example, signs, symbols and pictures.

- Use of ICT, such as Talk Buttons would be a great addition to this activity.

- Review which forms of communication the children liked best or were easiest to understand and why.

What the Ladybird Heard by Julia Donaldson

- Fact cards about ladybirds
- Plastic farm animals for sorting
- Sound makers that sound like the different animals or can be made to sound like them

- Sound map for children to follow relating to your school grounds (e.g. noisy children, chirping birds, squeaky gate)

Activity Card 1

The instructions tell the children to follow the sound map challenges children to work as a team: one blindfolded, one to lead the blindfolded child and the rest scatter around the area, and when the blindfolded child passes the children make the sound of one of the animals from the story – the blindfolded child has to guess what animal is making the sound.

10.6 Outdoors Science Story Sack – What the Ladybird Heard

Activity Card 2 (Personal Capability focus: Communication)

In a similar way to the story above, this book lends itself to reinforcing the need for and benefit of clear instructions. Develop the skill of direction and instruction giving using maps and directive words (right, left, straight ahead, before, etc.).

- On the playground set out an obstacle course that the children must direct a blindfolded friend through.

- In groups, role play the story within the school grounds, recreating the map and the instructions.

Duck in the Truck by Jez Alborough

- Fact cards about cars
- Toy cars and animals

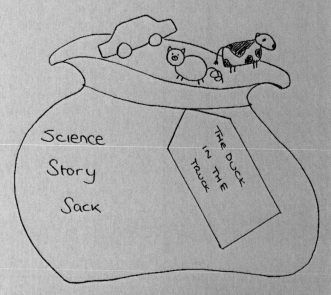

Activity Card 1

- What happens to car when you push it over different surfaces?

- How can you make it get stuck in some mud?

10.7 Outdoors Science Story Sack – Duck in the Truck

- What happens to the car when it goes down a hill? Does it go faster or slower?
- Pull a toy car out of wet sand or mud, using cotton or string.

Activity Card 2 (Personal Capability focus: Teamwork)
In this story the duck meets various other animals that all take time out from what they are doing to help him, sometimes not to their benefit. Such selfless behaviour demonstrates good teamwork.

- What kinds of words could we use to describe the animals that the duck met?
- What would you have done if the duck asked you for help?
- What words can you use to describe the way the duck behaved?
- If you were to choose three friends of your own to help you, who would they be and why?

Jasper's Beanstalk by Nick Butterworth

- Fact cards about seeds
- Fact cards about different type of beans (e.g. runner beans, French beans, mung beans, cannellini bean seeds)
- Packets of different bean seeds
- Trowel
- Seed pots (optional)
- Watering can
- Plastic plant labels and pencil
- Camera

10.8 *Outdoors Science Story Sack* – Jasper's Beanstalk

Activity Card 1
- Where will you plant your bean seeds in the school grounds?
- What will your bean seeds need to help them germinate?
- Which is the best place for the bean seeds to grow?
- How will you make sure that you and other people know where the seed has been planted?
- How will you show other people what kind of bean plant will grow from each seed?

Activity Card 2 (Personal Capability focus – Self-management)
In this story Jasper attempts to grow his beanstalk, but has not got the patience to give it the time it needs to shoot and grow.

- Ask the children what it means to be patient?

- What other words could we use instead of patient?

- When have they had to wait a long time for something? How did it feel?

- What tips could they give Jasper to help him be more patient or to stick at it?

Moonbear's Shadow by Frank Asch

In the Science Story Sack there should be a Fact Card about shadows and how shadows are made.

- Camera
- Bear
- Torch

Activity card 1
- How many different shadows can you find around the school grounds?

- Which is the biggest shadow you can find?

10.9 Outdoors Science Story Sack – Moonbear's Shadow

- Which is the smallest shadow you can find?

- Make an unusual shadow with a friend.

- Which is the strangest shadow you can find?

- Make Moonbear's shadow.

- How can you make Moonbear's shadow disappear?

- Be like Moonbear; how can you get rid of your shadow?

You could use the digital camera to photograph shadows

Activity Card 2 (Personal Capability focus – teamwork)
In this activity the children are asked to work with a friend to make different shadows, which means that they will have to work towards a common goal, listen to each other, and perhaps even change their own ideas. All these Personal Capabilities are important to working in science.

Egg Drop by Mini Grey

- Fact cards about parachutes
- Pictures of different parachutes
- Different parachutes with people or hard boiled eggs on the end
- Metal washers

Activity Card 1

- Which parachute is the best?

- How can you make a parachute fall more quickly?

- How can you make a parachute fall more slowly?

- What happens if you add metal washers to the parachute?

Activity Card 2
(Personal Capability focus: Creativity and Problem Solving)

In this story the egg has a dream, although his attempts may not prove as

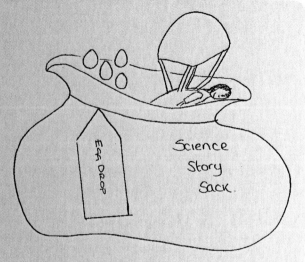

10.10 Outdoors Science Story Sack – Egg Drop

successful as he hoped. When the inevitable happens there are a range of attempts to put the Egg back together, from using string to tomato soup. Ask the children to explore the ways that a parachute could be damaged and, of course, how it could be fixed. Reinforce the benefit of not just opting for their first guess, but that exploring alternative ideas and joining up people's ideas is a useful creative problem-solving strategy.

Stick Man by Julia Donaldson and Axel Scheffler

- Fact cards about trees being used to make wooden objects

- Picture of everyday wooden objects

- Pictures of unusual wooden objects

- Collection of wooden objects

- Pieces of wood (e.g. balsa, twigs, craft sticks)

- Precut pieces of string

- Camera

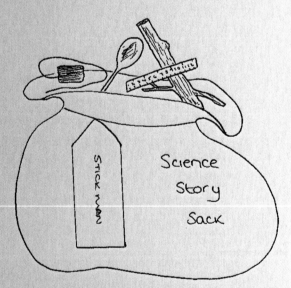

10.11 Outdoors Science Story Sack – Stick Man

Activity Card 1

- Go around the school and make a list or take photographs of things that are made from wood.

- Think about why wood is useful material, and why objects around the school are made from wood.

- Use the wood and the string in the sack and reuse them to make something useful.

Activity Card 2 (Personal Capability focus: Self-awareness)

In this story the Stick Man is desperate to go back to the Family Tree. Talk to the children about their families and who is in their family. If possible ask them to bring in pictures, or draw pictures, of their families and to look at the features of those people – eye colour, height, hair colour and so on.

- Create a Family Tree of their own by putting their own photographs/pictures of their families in plastic wallets with a piece of card to strengthen them (make sure parents are happy that the photographs may not be returned). Pin or hang them from trees in the school grounds. Making a tree personal in this way will relate their own circumstances to those in the story and can become a lovely talking point during playtimes and home times.

Reference

Feasey, R. (2006) Using stories and poems in science primary, *Science Review*, 92, March/April: 8–10.

CHAPTER

11 The future

We hope that this book has challenged you to think of ways forward in taking primary science 'Beyond the Classroom Boundaries' and laying the foundations of children's and staff relationships with their immediate environment – where the school grounds are science learning places.

Framwellgate Primary School in Durham has been doing exactly that, and their vision and way forward is shown below:

Our Vision:

Children will learn through having:

- A wealth of Outdoor Learning experiences which are an integral part of the curriculum.
- Teachers who embrace Outdoor Learning with enjoyment and confidence.
- An environment which provides stimulating and interactive learning.
- Resources which are easily accessible for independent learning.

Framwellgate's Vision for Outdoor Learning has been translated into a school policy document so that all those involved with the school understand why outdoor learning is such a priority for the whole school and why children spend so much time working in Science: Beyond the Classroom Boundaries.

The policy looks at:

- What is Outdoor Learning?
- Where can Outdoor Learning take place?
- Personal Development / Social Development
- Organization
- Curriculum Links
- Outdoor Learning and Play
- Children with additional needs (SEN/MAT)

- Health and safety
- Appendices
  - ➢ County recommendations – adult to children ratios
  - ➢ Risk assessment pro forma
  - ➢ Children's risk assessments pro forma.

This list will probably be familiar to you, although maybe there is one section that is not yet included in all such policies. In Framwellgate's list it is found in the Appendices and it is 'Children's risk assessments pro forma'. In essence, this section encapsulates the vision this school has of how children will use and behave in science in the school grounds. Framwellgate has put the child at the heart of science outdoors by sharing the responsibility for risk and safety with their young people.

Like many schools Framwellgate is not throwing away the statutory curriculum, just rethinking where the best place for children to access it is, and the outdoors offers greater potential than they had realized. So the future for staff and children is one where the barriers between outdoors and indoors have been taken down and experiences in science are no longer defined by four walls.

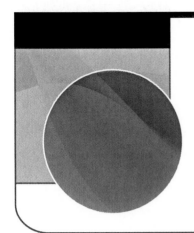

Index

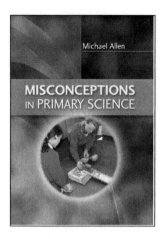

MISCONCEPTIONS IN PRIMARY SCIENCE

Michael Allen

9780335235889 (Paperback)
February 2010

This essential book offers friendly support and practical advice for dealing with the common misconceptions encountered in the primary science classroom.

Most pupils will arrive at the science lesson with previously formed ideas, based on prior reasoning or experience. However these ideas are often founded on common misconceptions, which if left unexplained can continue into adulthood.

This handy book offers 100 common misconceptions and advice for teachers on how to recognise and correct such misconceptions.

Key features:

- Examples from the entire range of QCA Scheme of Work topics for Key Stages 1 and 2
- Practical strategies to improve pupils' learning
- Support for teachers who want to improve their own scientific subject knowledge

www.openup.co.uk

 OPEN UNIVERSITY PRESS
McGraw - Hill Education

ESSENTIAL PRIMARY SCIENCE

Alan Cross and Adrian Bowden

9780335234615 (Paperback)
2009

This book offers you practical guiding principles which you can apply to every lesson. There are tips on how to ensure each lesson includes both practical and investigative elements and suggestions on how to make your lessons engaging, memorable and inclusive.

Each chapter is organized around the following structure:

- What science do you need to know and understand?
- What science do your pupils need to learn?
- What is the best way to teach these topics in the primary classroom at KS1 and KS2?

Sample pupil activities are also included and there is coverage of how to deal with common misconceptions within every chapter.

www.openup.co.uk

 **OPEN UNIVERSITY PRESS**
McGraw - Hill Education

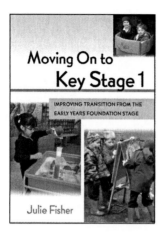

MOVING ONTO KEY STAGE 1
Improving Transition from the
Early Years Foundation Stage

Julie Fisher

9780335238460 (Paperback)
2010

eBook also available

The author considers recent evidence about how children learn and
questions whether current practice in Key Stage 1 optimises these
ways of learning. Challenging the reliance on teacher-directed
activity, she asks whether introducing more child-initiated learning
could offer children a more appropriate balance of learning
opportunities.

Key issues include:

- The place of play in Key Stage 1
- Organizing the learning day to include child-initiated activity
- Observation and assessment
- Planning
- The role of the teacher

www.openup.co.uk

OPEN UNIVERSITY PRESS
McGraw - Hill Education

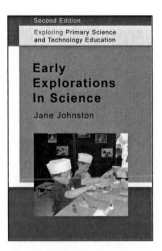

EARLY EXPLORATIONS IN SCIENCE 2/E

Jane Johnston

9780335214723 (Paperback)
2005

eBook also available

Early years' education has become firmly established as an important part of education. This 2nd edition celebrates good scientific practice in the early years, that is, with children up to 8 years of age. The book explores issues such as the range, nature and philosophical underpinning of early years' experiences and the development of emergent scientific skills, understandings and attitudes. It also considers how practitioners can develop creative scientific experiences and support children in their early scientific explorations.

New features for the second edition

- a focus on the development of early skills
- the importance of play in early development
- the importance of skilled interaction with well-trained adults
- the importance of informal learning (pre-school and home)
- the importance of a motivating learning environment

www.openup.co.uk

OPEN UNIVERSITY PRESS
McGraw - Hill Education